Environmental Economics and Investment Assessment II

WIT*PRESS*

WIT Press publishes leading books in Science and Technology.
Visit our website for the current list of titles.
www.witpress.com

WIT*eLibrary*

Home of the Transactions of the Wessex Institute.
Papers presented at Environmental Economics II are archived in the WIT eLibrary in volume 108 of WIT Transactions on Ecology and the Environment (ISSN 1743-3541).
The WIT eLibrary provides the international scientific community with immediate and permanent access to individual papers presented at WIT conferences.
Visit the WIT eLibrary at www.witpress.com.

SECOND INTERNATIONAL CONFERENCE ON ENVIRONMENTAL ECONOMICS AND INVESTMENT ASSESSMENT

Environmental Economics II

CONFERENCE CHAIRMEN

K. Aravossis
National Technical University of Athens, Greece

C. A. Brebbia
Wessex Institute of Technology, UK

N. Gomez
University of Cádiz, Spain

INTERNATIONAL SCIENTIFIC ADVISORY COMMITTEE

Organised by
Wessex Institute of Technology, UK
University of Cádiz, Spain
The National Technical University of Athens, Greece

Sponsored by
WIT Transactions on Ecology and the Environment

WIT Transactions

Transactions Editor

Carlos Brebbia
Wessex Institute of Technology
Ashurst Lodge, Ashurst
Southampton SO40 7AA, UK
Email: carlos@wessex.ac.uk

Editorial Board

Environmental Economics and Investment Assessment II

WIT PRESS Southampton, Boston

K. Aravossis
National Technical University of Athens, Greece

C. A. Brebbia
Wessex Institute of Technology, UK

N. Gomez
University of Cádiz, Spain

Published by

WIT Press
Ashurst Lodge, Ashurst, Southampton, SO40 7AA, UK
Tel: 44 (0) 238 029 3223; Fax: 44 (0) 238 029 2853
E-Mail: witpress@witpress.com
http://www.witpress.com

For USA, Canada and Mexico

Computational Mechanics Inc
25 Bridge Street, Billerica, MA 01821, USA
Tel: 978 667 5841; Fax: 978 667 7582
E-Mail: infousa@witpress.com
http://www.witpress.com

British Library Cataloguing-in-Publication Data

A Catalogue record for this book is available
from the British Library

ISBN: 978-1-84564-112-2
ISSN: 1746-448X (print)
ISSN: 1743-3541 (online)

The texts of the papers in this volume were set individually by the authors or under their supervision. Only minor corrections to the text may have been carried out by the publisher.

Printed in Great Britain by Athenaeum Press Ltd

Preface

This book contains the papers presented at the Second International Conference on Environmental Economics and Investment Assessment, held in Cádiz, Spain, in 2008. The conference was organised by the Wessex Institute of Technology in collaboration with the University of Cadiz and the National Technical University of Athens.

The current emphasis on sustainable development is a consequence of the general awareness of the need to solve numerous environmental problems resulting from our modern society. This has resulted in the necessity to assess the impact of economic investments on the environment. The book addresses the topic of Investment Assessment and Environmental Economics in an integrated way; in accordance with the principles of sustainability considering the social and environmental impacts of new investments.

The usual assumption is that it is difficult to achieve the growth of enterprise in an environmentally friendly manner. This paradigm usually associated with developed countries is now affecting all other regions of the globe. The main question is if the development of enterprise is compatible with environmental protection.

The roots of financial development are financial growth, which in conventional terms requires an increase in production and the use of more resources. Overuse of those resources can result in the destruction of the natural resources and a larger release of waste and pollution into the environment.

The book addresses these problems of primary importance to Society, discussing and proposing a more constructive and progressive approach to ensure sustainability. Methodologies to address these important problems are presented.

The contributions comprise the following broad subject headings: Economy and the Environment; Environmental Policies, Planning and Assessment; Cost–Benefits Analysis; Natural Resources Management; Environmental Performance; Social Issues and Environmental Policies.

This book will be of interest to government officials, politicians, environmental experts, economists, research scientists in the area of environmental economics, operations researchers, senior management in all kinds of companies and regional government.

The Editors are grateful to all the authors for their excellent contributions and in particular to the members of the International Scientific Advisory committee for their reviews of the abstracts and the papers and their help on ensuring the high quality of this book.

The Editors
Cádiz, 2008

Contents

Section 3: Cost benefits analysis

Section 4: Natural resources management

Section 5: Environmental performance

Section 6: Social issues and environmental policies

Section 1
Economy and the environment

Decision Aid Framework For Investments in Environment (DAFFIE). Investing in the environment; when is it too much?

S. Vanassche, L. Vranken & P. Vercaemst
VITO, Flemish Institute for Technological Research, Belgium

Abstract

DAFFIE provides a methodology to estimate the economic feasibility of environmental investments. Typically, (mandatory) environmental investments are unprofitable as such so that classical investment analysis provides insufficient insights to assess their economic feasibility. Therefore, we evaluate whether an industry or company has the carrying capacity for extra costs associated with the introduction of environmental measures. In the DAFFIE evaluation method, an investment is economically feasible when an industry or company is able to maintain or strengthen its competitiveness which is defined as the ability to maintain sufficient liquidity and solvency and to earn a return from activities that exceeds the cost of capital in the long run. This implies the use of financial ratio analysis. DAFFIE starts from the annual accounts of an (average) company. The impact of the investment options is simulated into projected accounts. Eight key financial ratios are calculated for the projected statement with the investment options included and excluded. Finally, the evolution of the key ratios is benchmarked against a reference group (e.g. total industry in Flanders) to come to a conclusion on the feasibility of the investment options. This methodology highlights the potential impact of unprofitable environmental investments on the financial position of a company or industry. The results are used to support environmental policy makers or to objectify discussions between companies which are confronted with additional investments due to environmental legislation, and governments which issue emission permits. The paper describes the DAFFIE methodology in more detail and illustrates its use with practical cases.

Keywords: environmental investment, financial ratio analysis, economical feasibility, calculation model.

WIT Transactions on Ecology and the Environment, Vol 108, © 2008 WIT Press
www.witpress.com, ISSN 1743-3541 (on-line)
doi:10.2495/EEIA080011

1 Introduction

By definition, environmental investments are implemented with the purpose of protecting the environment, often under pressure of environmental policy (e.g. permit conditions) [5, 6]. As a consequence, evaluating this kind of mandatory investments by 'classical' investment analysis, such as the payback time, the net present value or the internal rate of return provides insufficient insights in the discussion between industries and authorities. Rather than evaluating the (un)profitability of these investments, our aim is to evaluate whether an industry or company has the carrying capacity for the extra costs associated with environmental investments.

In the DAFFIE evaluation method, an environmental investment is evaluated as economically feasible when an industry/a company is able to maintain or strengthen its competitiveness in the short and in the long run. Competitiveness is hereby described as the ability to maintain sufficient liquidity and solvency and to earn a return from activities that exceeds the cost of capital in the long run [4]. This definition implies the use of financial ratio analysis.

In this paper, we first highlight the merits of DAFFIE compared to existing methods for evaluating environmental investments. Second, the method itself including the main calculation steps is elaborated. Finally, the role of DAFFIE in the decision process concerning environmental investments is clarified.

2 Merits of DAFFIE

Quite surprisingly, there is hardly any literature on the evaluation of environmental investments [7]. However, a number of practical methods exist such as Reference Values [4] an MIOW$^+$ [3]. While these methods provide a suitable base for the evaluation of the economic feasibility of environmental investments, they do entail some important disadvantages and shortcomings which are the main reason for developing a new decision aid tool.

First, the method of Reference Values [4] implies calculating the proportion between the costs of the environmental investment and a number of financial parameters. The proportion between the yearly net costs of the investment and turnover, gross added value and operating profit of the sector or company over four years are determined. In addition the share of the environmental investment of the total investments is determined. These proportions are then related to cut-off points, which allow classifying each of the proportions as 'unacceptable', 'further discussion needed' or 'acceptable'. The main shortcoming of this approach is that a large number of environmental investments fall within the range where further discussion is needed and therefore it provides no conclusive judgement about the economic feasibility of different types of environmental investments.

Second, MIOW$^+$ [3] is a model that allows one to estimate the consequences of an environmental investment for the economic situation of an individual company. It takes the financial situation in to account as well as the company's competitive position. The main disadvantage of this model is that it is designed for evaluation of investments for large individual companies. Hence, it is not suitable to evaluate the viability of environmental investments for entire industries or SME's.

DAFFIE has the advantage that it can be applied to an entire industry as well as to an individual company and to large as well as small- and medium-sized enterprises. In addition, DAFFIE provides a more conclusive outcome for all types of environmental investments. Furthermore, DAFFIE benchmarks each firm or industry against a reference group and uses this relative position, rather than absolute boundaries for financial ratios, to determine the economic feasibility of environmental investments.

3 Methodology

3.1 DAFFIE framework captured in an Excel-based tool

Figure 1 shows the general overview of the steps taken within the DAFFIE framework.

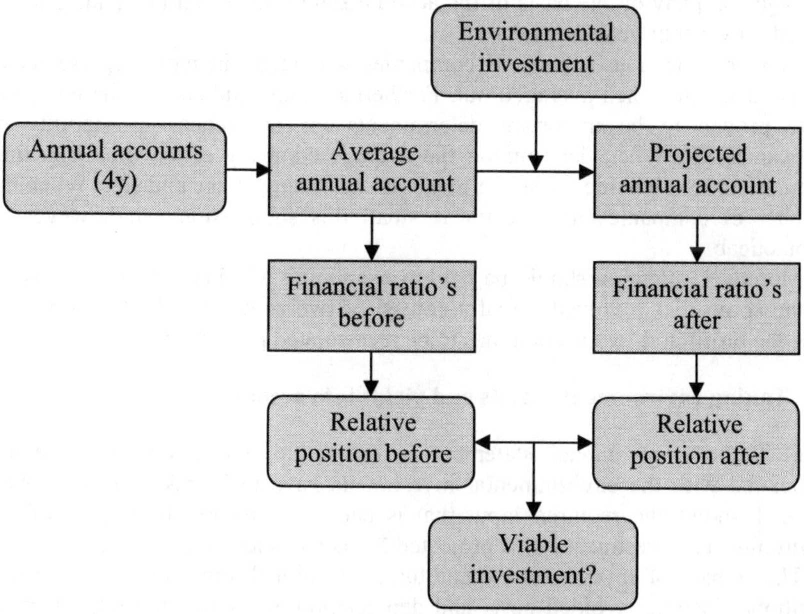

Figure 1: Calculation steps within the DAFFIE framework.

First, the necessary financial data is extracted from the company's annual accounts of the four latest years available. This information is used to construct. The accounts are averaged out over four years in order to flatten out yearly fluctuations into an 'average annual account'. Next, input data concerning the environmental investment options is needed in order to simulate a projected annual account on the basis of the average annual account taking the net costs of the environmental investment options into account. Financial ratios are then calculated for the actual as well as the projected annual account. In a final step the relative position of all calculated ratios is compared to the ratios from a reference group (e.g. food industry in Flanders).

The comparison between the relative position of the ratios before and after the environmental investment then serves as an objective starting point for the discussion of the viability of the investment in question.

In order to ease these extensive calculations, we designed a tool as an Excel-workbook containing a number of input fields, calculation sheets and output fields. This tool is in first instance designed to fit the Belgian accounting system as it automatically extract the needed financial data from a database containing all deposited annual accounts of Belgian companies.

3.2 Defining an average company

DAFFIE can be used not only for analysis at the level of an individual company, but also at the level of an industry, for example to determine sectoral Best Available Techniques. The annual account then is drawn up for an artificial average company on the basis of the account statements of all companies in the industry over four years considered.

More homogenous groups of companies will result in more representative annual accounts when averaged out. Furthermore, age and size of the company have proven to be important determinants for differences in viability of companies [1]. When determining the average company of a sector it is thus appropriate to subdivide the sector in classes according to age and size. When the number of companies in a sector is small this subdivision can however be impracticable.

Moreover, attention should be paid to examining whether the financial years of the companies involved actually consist of twelve months. If this is not the case the profit-and-loss account has to be recalculated proportionally.

3.3 Taking environmental costs and yields into account

Once the average annual statement is drawn up, the costs and revenues associated with the environmental investments have to be taken into account. Table 1 shows the required input that is needed to record the impact of the environmental investment into a projected financial statement.

The impact of investment expenditure, operational and maintenance costs, additional income, avoided costs and depreciation as well as the impact of an additional loan on the balance sheet and profit-and-loss account are all taken into account.

Certain assumptions need to be made regarding the amounts involved, the way it is financed, the depreciation rate and taxes. These can be easily be adjusted in different scenarios.

On the basis of the average and projected accounts, financial ratios are calculated representing respectively the financial health of the sector or company before and after the implementation of the investment.

Table 1: Input cost components for DAFFIE.

Cost component	Unit
Investment amount	
Expenditure on pollution control equipment	k€
Installation expenditure	k€
Additional parameters	
Economic life span	years
Depreciation Installation Expenditure	yes/ no
Financed with own funds	%
Term of debt	years
Interest rate of debt	%
Operating and maintenance cots	
Yearly personnel costs	k€
Other yearly operating and maintenance costs	k€
Additional income and avoided costs	
Yearly income	k€
Yearly avoided costs	k€
Other	
Corporate tax rate	%

Ooghe et al [2] developed an intuitive failure prediction model based on the eight financial ratios represented in Table 2. The four major areas of financial health, added value (AV), profitability (P), solvability (S) and liquidity (L) are covered by this group of ratios. The average of the logit values of these ratios forms the FiTo®-score which is an indicator of the general financial health.

Table 2: Financial ratios used to assess financial health.

Ratio	
1. Gross added value / personnel costs	AV/P
2. Net return on operating assets before taxes	P
3. Net return on shareholders funds after taxes	P
4. Self financing quote	P/S
5. Financial independence ratio	S
6. Short term financial debt ratio	S
7. Coverage of external liabilities by cash flow	R/S
8. Net treasury ratio	L

The eight financial ratios and their corresponding FiTo®-score are expressed as a percentile score with respect to a reference industry (e.g. total industry in Belgium). This allows quantification of the influence of the investment on the financial situation.

4 Decision framework

Figure 2 visually presents the financial situation before and the estimated financial situation after the implementation of an environmental investment. The eight numbered axes each depict one financial ratio in accordance with the numbering in table 2. The ratios are represented as percentile values with respect to the reference industry 'total industry in Flanders'. For instance a percentile value of 60 for ratio 1 means that the average company from the sector in question has a larger gross value added over personnel costs than 60% of the companies from the reference group.

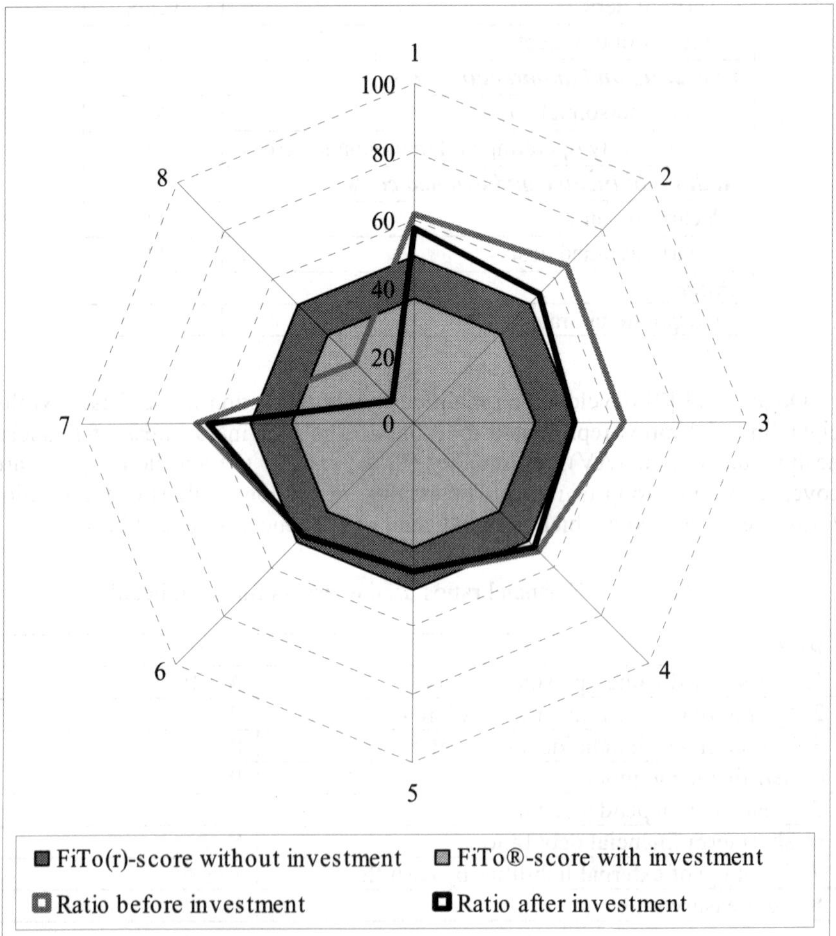

Figure 2: Visualisation of the financial impact of an environmental investment.

The shaded areas in the diagram in figure 2 represent the percentile position of the FiTo®-score [2] indicating the general financial health of the average company within the sector in question. Corresponding with the financial ratios, the FiTo®-score is also depicted for an average company before the investment (light grey area) and for the estimated average company after the environmental investment (dark grey area).

The financial ratio analysis and FiTo®-score intuitively provide two manners of evaluating the viability of an environmental investment. In the first manner the financial ratios are examined individually or within their area of financial health; added value, profitability, solvability and liquidity. When at least one ratio worsens unacceptably it can be stated that the environmental investment is not viable. In the second manner of evaluation, the general financial health is taken into account by assessing the worsening of the FiTo®-score by the implementation of an environmental investment. Both manners can complement each other. Further investigation and application of the framework to a number of industry case-studies will help to clarify which worsening of financial ratios and FiTo®-score are acceptable for an economic feasible investment.

5 Conclusion

The DAFFIE methodology provides valuable insight in the possible effects of (unprofitable) environmental investments on the financial position of a company or industry. This information can be used to objectify discussions between companies confronted with environmental legislation and governments for environmental permitting cases or as policy support.

In order to ease calculations a calculation model is developed for the Belgian situation. This model can be adapted to meet accounting specifications in different regions or to contain other financial ratios.

At the moment a number of industry case studies are being carried out in order to validate the model and provide more insight in the evaluation of the projected financial situation after the implementation of the investment. Further attention is also paid to establishing the limiting conditions for the application of DAFFIE.

References

[1] H. Ooghe, C. Spaenjers, P. Vandemoere, *The Financial state of the Belgian companies 2005*, Intersentia, 2005, Gent.

[2] H. Ooghe, C. Spaenjers, *The financial state of the Belgian companies 2006; Ratios and total score on the basis of the FiTo®-meter 1995-2006,* Department of Accountancy and Corporate Finance Ghent University Working Paper 2006/380, April 2006, Gent, 22 p.

[3] K.F. Van der Woerd, I.A.W van Rijn., S. Rosdorff, E. Masurel, Y.M. van Everdingen en R.B. Dellink, *MIOW$^+$; background to the model*, Institute for Environmental Studies, Vrije Universiteit Amsterdam, 1995.

[4] P. Vercaemst, *BAT: When do Best Available Techniques become Barely Affordable Technology?*, BAT-centre VITO, May 2002, Mol.
[5] VROM, *Kosten en baten in het milieubeleid: definities en berekeningsmethoden,* 1998, Den Haag, 554 p.
[6] LNE, *Milieubeleidskosten – begrippen en berekeningsmethoden*, Brussel, 41 p.
[7] European IPPC Bureau, *Reference document on economics and cross-media effects,* European Commission, July 2006, Sevilla.

Value and price: a transdisciplinary approach to urban water management

K. Kviberg
School of Geography Geology and Environmental Science,
University of Auckland, New Zealand

Abstract

This paper puts forward the benefits of a transdisciplinary approach to environmental management, using urban New Zealand use and value of freshwater as a case study. New Zealand is not water poor, but suffers from regional seasonal stresses on freshwater resources. The old mindset that water is a free gift from nature prevails within the water management industry, encouraging unsustainable consumption behaviours that are likely to develop into significant economic dis-benefits. Existing pricing structures are obscuring signals of both water shortages and wasteful practices, raising concerns that New Zealand urban water is underpriced and undervalued. The paper aims to fit a transdisciplinary research framework, drawing from economic, social and environmental disciplines, to establish options for resource management and asset investment. By the coupling of values with consumption, and the willingness to pay for environmental goods and services, this framework maximises environmental and welfare objectives aligned with sustainable development goals.

Keywords: choice modeling, low-impact urban design, pricing, transdisciplinarity, values, water management.

1 Introduction

Improved understanding of the interconnectedness in nature and the complexities of environmental problems emerged at an increasing rate throughout the second half of the last century [1–3]. The deficiencies of discipline-based research in dealing with these problems has become increasingly evident, resulting in the

www.witpress.com, ISSN 1743-3541 (on-line)
doi:10.2495/EEIA080021

coordination of multi-disciplinary research teams, inter-disciplinary research projects, as well as the transdisciplinary approach [4–8].

It has been suggested that the uptake of transdisciplinary research will better address the complex socio-environmental problems currently undermining sustainable development efforts [4, 6, 9]. Transdisciplinary research is not antagonistic but complimentary to disciplinary, multidisciplinary and interdisciplinary research. It is seeking to understand the workings of whole systems as they exist, as opposed to producing knowledge of their constituent parts, which is aptly provided for by each discipline [6]. In accordance with systems theory, it views a system as being more than its constituent parts, and that sustainable solutions come from the synergy of disciplinary methods. A transdisciplinary framework needs to be cost-effective, socially acceptable, with transparent responsibilities and accountabilities. Transdisciplinary activities include problem definition, problem representation and problem solving [9].

Water is a perfect example of a sustainable development challenge encompassing environmental, economic and social dimensions. Reconciling these three aspects is a significant policy challenge for governments [10]. Water is a unique raw material, essential for life, economic activity, and cultural identity. Global trends show a quadrupling in water demand due to industrialisation and irrigation and a decline in available water supplies by 40% since 1970; an increase in costs relating to more distant and poorer water quality supplies; with a corresponding increase in energy consumption to meet water demands [11, 12]. Freshwater, once considered the ultimate renewable resource, is currently utilised in terms of both take and degradation at a rate exceeding the rate of natural replenishment on a global scale. The old mindset that water is a free gift from nature continues to encourage squandering of the resource. Current management of freshwater resources in New Zealand cities is unsustainable and likely to develop into significant economic dis-benefits on regional and national scales [13].

This paper posits that a transdisciplinary approach may assist management practices better aligned with sustainable development goals. It is structured under the main components of transdisciplinary research: problem definition; problem representation and problem solving.

2 Problem definition: current management practice

Originating from the experience of health benefits as a result of improved sanitation in the mid to late 1800's, the "big-pipe-in big-pipe-out" centralised water management strategy has been the norm in almost all industrialised cities for the last 150-200 years [14]. The designs of these systems reflect the general paradigm related to natural resources of the time; considering water a gift from nature and assuming free disposal of polluted water to the environment. With urban centres reaching the size they are today and with continued rapid growth; this approach has become a classic example of Hardins' "tragedy of the commons" [15]; the over-extraction in water-scarce regions, and the degradation of urban streams and harbours in wetter regions. Community perceptions and the willingness to accept policy change is poorly understood in most water management districts. This commonly prohibits the implementation of technological and policy improvements, affecting asset investment decisions and thus continues to foil attempts to create sustainable, low impact urban communities.

2.1 Social

How resources are managed depends upon a society's perception of value. This may change with an improved understanding of management options and subsequent consequences for the economy, society or the environment. Thus, imperative to sustainability is the continuous evaluation of community attitudes and expectations with regards to resource management, and mapping of divergent interests to deal appropriately with stakeholder conflicts [16]. Only by knowing these multiple realities can one design policies that, whilst deemed acceptable, pull communities' behaviours away from wasteful consumption towards ecologically sustainable resource utilisation and asset investment.

The implementation of large scale demand management strategies have consistently demonstrated that technological fixes alone can not adequately address issues of urban freshwater management [17–21]. Evidence is emerging that there is support for demand management to be utilised to its full potential, and that pricing structures/adjustments are favoured over water rationing both in New Zealand and other countries [22–27].

Effecting behavioural changes demand that stakeholders (or problem owners/ water consumers) take responsibility for the environmental effects of their actions. Kolokytha *et al* [18] and Nancarrow *et al* [19] explored the relationship between how people perceive themselves as water consumers and consumption behaviour. Both studies found that how consumers valued water did not significantly affect consumption behaviour. These results indicate a present decoupling of values and environmental behaviours.

2.2 Economic

The traditional economic approach to infrastructure development considers water a public good funded in full by governmental agencies. However, shifting trends in water pricing in OECD countries have seen an increase in the use of full cost-recovery pricing structures, and an increasing use of volume based charges [10]. It is considered that this shift has contributed to a stabilisation of demand, or in some European cities, net reductions in demand [28]. Urban water supply contains two major components of cost: infrastructure and storage, and supply operations. The cost recovery process likewise entails two components, one to cover the investment in infrastructure assets, and the other the marginal cost of water use [24, 29]. Scarcity and/or ecological externalities on the other hand are commonly not factored into the pricing equation. Charges for water are typically set by local utility operators owned or subsidised by local government and are charged either through the general rating structures, or by volume, with an additional fixed charge component. Such pricing structures for urban water supplies prevent signals of water shortages and encourage wasteful practices raising concerns that urban water is both under-priced and under-valued, with cross-subsidies commonly obscuring signals of the real value of water [13].

Social equity concerns are often, and justifiably, contentiously argued in the water pricing debates. The established benchmark affordability level for household water and sanitation expenditure is set at 4-5% of the household income [10]. On average, residents in New Zealand cities spend 2% or less for water and sanitation, and thus can afford to squander the resource.

Allocation of water resources is rapidly becoming a significant issue in Christchurch, New Zealand's third largest urban centre, where potable supplies and urban streams both originate from high quality aquifers. Abstraction for urban supplies has caused diminished water-flows in streams, affecting both ecological and recreational values. The community's willingness to pay for the retention of flows in Christchurch streams has been found to be relatively high, suggesting that some New Zealand communities readily accept that environmental protection has benefits [25]. Whether this is reflected in a willingness to change behaviour, or merely suggesting an acceptance of importing water from rivers further a field has not been subjected to research.

2.3 Cultural

Water is a unique raw material, not only essential for life and economic activity, but also people's cultural identity. Recreation, amenity values and the spiritual connections to water is by current management practise decoupled from people's perceptions of the value and willingness to pay for water in the city.

The New Zealand cultural perspective is unique. Maori, New Zealand's indigenous people, proclaim a special relationship with water and have inherent strong 'views' on, and are sensitive to, water management practises and the protection of mauri (life-force) [13]. It has been suggested that to meet sustainability objectives, infrastructure services need be provided in a way that protects cultural heritage values, give due consideration to customary interests in natural resource use, and allow participation of Maori as partners in decisions regarding natural resources [30].

3 Problem representation: segregation

How do these aspects interlink into one system? For urban water management to be sustainable it must be integrated across management fields and consider potable water, stormwater, wastewater and natural water bodies as part of one system. This approach requires the integration of built form into the strategic planning, as well as demand management and water recycling schemes across several scales [31]. Barriers to such integration remain amongst water industry managers, policy makers and the consumer community at large.

Figure 1 shows a segregated, linear urban water management system. Stormwater, potable water and natural water bodies are commonly managed by isolated entities. The clear boxes show components where New Zealand water management is currently lacking in technological and/or policy implementation and asset investment. Thus, these *inactive* elements do not influence price, nor do they allow pricing to be an instrument for demand management as a response to physical stocks.

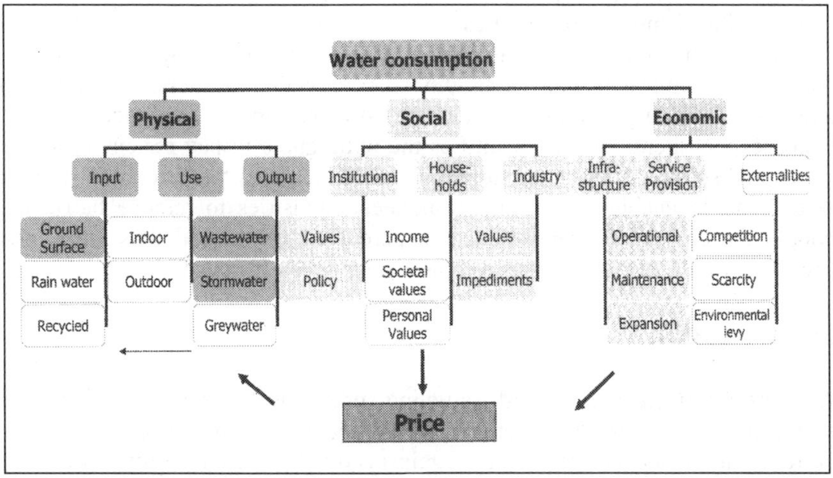

Figure 1: Urban water consumption.

4 Problem solving: coupling values and price

This paper purports addressing unsustainable urban water management trends through a framework that draws from the disciplines of urban ecology, resource economics, ecological economics, engineering, water industry and planning/policy science from a values perspective. Figure 2 shows this transdisciplinary framework for the urban water management sector including components at four cognitive levels, and illustrates the broad range of management fields that need to be included (as opposed to consulted with) as stakeholders in a holistic water management system.

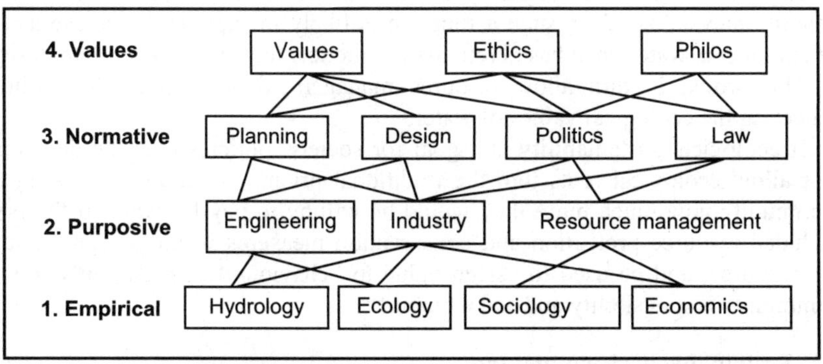

Figure 2: A transdisciplinary framework for urban water management (Adapted from [5]).

The role of a transdisciplinary researcher would be to facilitate the inclusion of methodologies from these fields, to map divergent interests, to promote synergy and learning between and within each participant, and to devolve responsibilities and accountabilities.

The construction of a transdisciplinary framework that 1) identifies consumer barriers to technical fixes such as greywater recycling and rainwater collection and use (reducing polluted runoff to streams and stormwater treatment expenses); 2) identifies and quantifies the willingness to pay for environmental goods and services (consumer surplus); and 3) identifies the perceived values of water; should enable the water management industries to remove barriers, to couple the consumer surplus with the perceived values and thus allow asset investment to be ecologically as well as economically efficient.

5 Discussion

A transdisciplinary framework aligning urban water pricing policy with sustainable development objectives should include performance indicators of current water management practises against ecological sustainability indicators; a pricing structure for urban water that promotes sustainable resource use; and a strategy for enhancing public perception and value of water to encourage the acceptance of a pricing structure supporting desirable economic, environmental and social outcomes.

5.1 Value and price

The discrepancy between communities' values and perceptions on one hand and urban water consumption behaviour on the other can be explained by the decoupling of water as an ecological good, and water as an economic good (Figure 3). Analyses of consumers' understanding of pricing policies and their willingness-to-pay response to alternative pricing structures could clear the way for incorporating an extended subjective utility function accounting for ecological values and ecologically sustainable water service provision. A cost-benefit analysis based on such a function is likely to support the investment in decentralised, water-shed based reticulation models for low impact urban design.

This would in turn allow resource managers to develop pricing policies reflecting the ecological values of water.

If ecological sustainability is a goal for society, policies must be developed that allow ecological rents into the traditional pricing equation. By asking the community how much more they would be willing to pay for water if the price included resource protection and conservation measures would assign value to the resource, and be used to assign value to behavioural changes, encouraging community responsibility and accountability.

5.2 Sustainable decision making

Conventional cost-benefit analysis (CBA) allows choosing a policy alternative among others on the criterion of economic efficiency. The use of multi-criteria analysis (MCA) has increased significantly in the last two decades as an attempt to satisfy often conflicting objectives of multiple stakeholders in decision making [32]. Holz *et al* [33] point to some of the limitations of CBAs and MCAs, and posit bridging the dichotomous "calculate or communicate" divide by the structure of a framework drawn from both approaches.

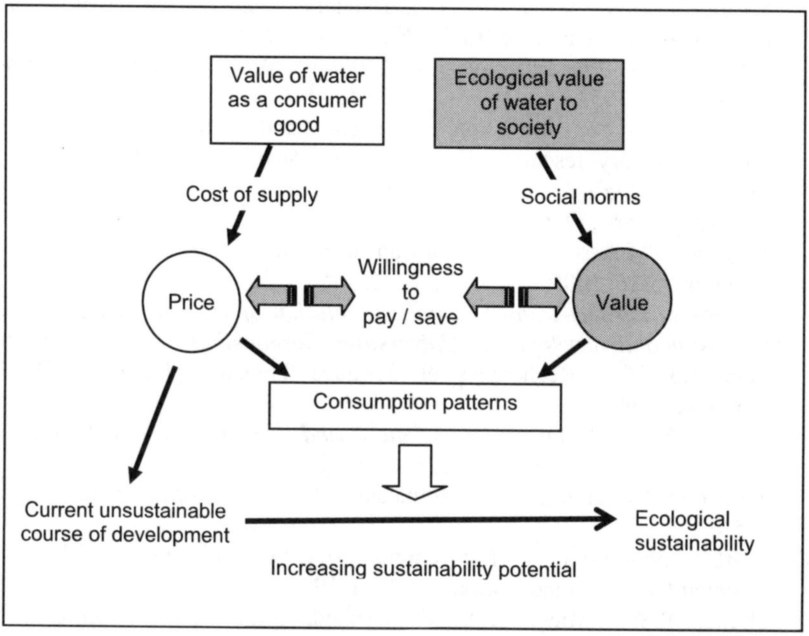

Figure 3: Relationship between values, price and consumer behaviour.

One approach to policy analysis in a transdisciplinary framework is the use of system dynamics modeling. System dynamics models the interactions of population, ecological and economic systems using feedback loops [34]. Stocks and flows connections are used to run "what if" simulations to test certain policies, and can as such greatly aid managers and communities in understanding how the system changes over time [35, 36].

Effecting behavioural changes demand that consumers take responsibility for the environmental effects of their actions. This needs to be facilitated by policies aimed at encouraging environmentally responsible behaviour. The benefits of using a transdisciplinary approach to assess current and alternative water management, investment and pricing policies include the capacity to couple choice modeling with system dynamic modeling, allowing consumers to visualise the consequences of management options aiding in assigning value to behavioural changes.

References

[1] Ehrlich, P.R. & Ehrlich, A.H., Why isn't everyone as scared as we are? (Chapter 1). *Valuing the earth: economics, ecology, ethics*, eds. H.E. Daly & K.R. Townsend, Massachusetts Institute of Technology Press: Cambridge, pp. 55-67, 1993.

[2] Meadows, D.H., Meadows, D.L., Randers, J. & Behrens, I.W.W., *The limits to growth*, Universe Books: New York, 1972.

[3] Odum, E.P., *Fundamentals of Ecology*, W. B. Saunders, Comp.: Philadelphia, 1953.

[4] Becker, E., Transformation of social and ecological issues into transdisciplinary research. *Knowledge for Sustainable Development: An Insight into the encyclopaedia of life support systems,* UNESCO / Eolss Publishers: Paris, Oxford, 2002.

[5] Max-Neef, M.A., Foundations of transdisciplinarity. *Ecological economics*, **53(1)**, pp. 5-16, 2005.

[6] Nicolescu, B., *The transdisciplinary evolution of the university condition for sustainable development, Universities' Responsibility to Society*, Centre International de Recherches et d'Etudes Transdisciplinaires (CIRET): Bangkok, 1997.

[7] UNESCO, *Conference on science and tradition: transdisciplinary perspectives on the way to the 21st Century-1991*, www.transdcongress.com.br/interna.asp?idCliente=83&acao=materia&id=5850

[8] World Commission on Environment and Development (WCED), *Our common future*, United Nations: Oxford, 1987.

[9] Sholtz, R.W., Mieg, H.A. & Oswald, J.E., Transdisciplinarity in groundwater management- towards mutual learning of science and society. *Water, Air, and Soil Pollution*, **123**, pp. 477-487, 2000.

[10] OECD, *Water: Performance and challenges in OECD countries*, OECD: Paris, 2003.

[11] OECD, *Improving water management. Recent OECD experience*, OECD / IWA: Paris, 2003.

[12] World Water Assessment Programme. *Valuing Water*, www.unesco.org/**water**/wwap/

[13] PCE, *Ageing pipes and murky waters. Urban water system issues for the 21st Century*, Parliamentary Commissioner for the Environment: Wellington, 2000.

[14] Livingston, D.J., Stenekes, N., Colebatch, H.K., Ashbolt, N.J. & Waite, T.D., Water recycling and decentralised management: the policy and organisational challenges for innovative approaches. *Conference proceedings from WSUD 2004*: Adelaide, 2004.

[15] Hardin, G., Tragedy of the commons. *Science*, **162**, pp. 1243-1248, 1968.

[16] Tacconi, L., *Biodiversity and ecological economics: participatory approaches to resources management*, London: Earthscan Publications Ltd, 2000.

[17] Burn, L., De Silva, D. & Shipton, R., Effect of demand management and system operation on potable water infrastructure cost. *Urban Water*, **4**, pp. 229-236, 2002.
[18] Kolokytha, E., Mylopoulus, Y. & Mentes, A., Evaluation demand management aspects of urban water policy- A field survey in the city of Thessaloniki, Greece. *Urban Water*, **4**, pp. 392-400, 2002.
[19] Nancarrow, B.E., Po, M., & Porter, N.B., Why doesn't it work? The incorporation of community culture in the evaluation of Water Sensitive Urban Designs. *Conference proceedings from WSUD 2004*: Adelaide, 2004.
[20] Nauges, C., & Thomas, A., Long-run study of residential water consumption (Chapter 2). *Current issues in the economics of water resource management*, eds. P. Pashardes, T. Swanson & A. Xepapadeas, Kluwer Academic Publishers: Dordrecht, pp. 47-66, 2002.
[21] Thomas, J.F. & Syme, G.J., Estimating residential price elasticity of demand for water: a contingent valuation approach. *Water Resources Research*, **24(11)**, pp. 1847-1857, 1988.
[22] Chapman, R., Goldberg, E., Salmon, G. & Sinner, J., *Sustainable development and infrastructure*, The Ministry for Economic Development: Wellington, 2003.
[23] Craig, J., Science and sustainable development in New Zealand. *Journal of the Royal Society of New Zealand*, **34(1),** 2004.
[24] Garcia, S. & Reynard, A., Estimating the benefits of efficient water pricing in France. *Resource and Energy Economics*, **26**, pp. 1-25, 2004.
[25] Kerr, G., Sharp, B. & White, P., The economics of augmenting Christchurch's water supply. *Journal of Hydrology (NZ)*, **42(2)**, pp. 113-124, 2003.
[26] MfE, *Sustainable development for New Zealand. Programme of action*, Ministry for the Environment: Wellington, 2003.
[27] Quiggin, J., *Environmental Economics and the Murray-Darling river system*, Faculty of Economics and Commerce, Australian National University, 2000.
[28] OECD, *The price of water: trends in OECD countries*, OECD, 2000.
[29] O'Fallon, C., *Linkages between infrastructure and economic growth*, Ministry for Economic Development: Wellington, 2003.
[30] NZIER, *Sustainable infrastructure: a policy framework*, NZIER: Wellington, 2004.
[31] McLean, J., Aurora- delivering a sustainable urban water system for a new suburb. *Conference proceedings from WSUD 2004*: Adelaide, 2004.
[32] Raju, K.S., Duckstein, L. & Arondel, C., Multicriterion analysis for sustainable water resources planning: a case study in Spain. *Water Resources Management*, **14**, pp. 435-456, 2001.
[33] Holz, L., Kuczera, G.& Kalma, J., Sustainable urban water planning in Australia: a decision science perspective. *Conference proceedings from WSUD 2004*: Adelaide, 2004.

[34] Forrester, J.W., Counterintuitive behaviour of system dynamics. *Technology Review*, **73(3)**, pp. 52-68, 1995.
[35] Constanza, R. & Ruth, M., Modeling for scoping, research and management (Chapter 2). *Institutions, ecosystems, and sustainability*, eds. R. Constanza, B.S. Low, E. Ostrom & J. Wilson J, CRC press LLC: Boca Raton, Florida, pp. 169-178, 2001.
[36] Richmond, B. Systems dynamics/systems thinking: let's just get on with it. *International Systems Dynamics Conference*: Stirling, Scotland, 1994.

Tariff regulation of the integrated water service: an Italian case

F. Cucchiella, M. Gastaldi & D. Frigioni
Department of Electric and Information Engineering,
Faculty of Engineering, University of L'Aquila,
Monteluco di Roio, 67040 L'Aquila, Italy

Abstract

In the water sector the tariff represents the regulatory service instrument through which the process of convergence between the conditions offered by providers and the conditions desired by collectivity is activated. At the same time the tariff must cover the cost of production and management of the service that overcomes the logic, sustained for far too long in Italy, to furnish an essential resource free of charge. The purpose of this work is to present the economic technical plan of an Optimal Territorial Basin (OTB) for an Italian city (L'Aquila) and the determination of the tariff adopted by the Integrated Water Service (IWS) in the next thirty years. Moreover this plan is able to activate forms of verification and incentive of the technical and economic efficiency that are the basis of water industrial management.
Keywords: price cap regulation, water management, network.

1 Introduction

The water sector provides an essential service of public utility; therefore it is necessary to guarantee the continuity of the service supply and an elevated qualitative standard to an accessible tariff. It's essential, therefore, that the management of the water resource is efficient and that a suitable system of regulation is used. Law 36/94 "dispositions in subject of water resources" (Law Galli) introduced a new way regarding the management of water services to overcome its fragmentation which is argued as the principal cause of the diseconomies and the dysfunctions of the sector (Barbese and Meucci [1], Caselli and Peruzzi [2]); for water services, it is meant here all those services

WIT Transactions on Ecology and the Environment, Vol 108, © 2008 WIT Press
www.witpress.com, ISSN 1743-3541 (on-line)
doi:10.2495/EEIA080031

linked with what Law 36/94 calls Integrated Water Service (IWS) i.e. those dealing with supplying – fetching, transporting and distributing – water for domestic use, as well as those dealing with the collection and treatment of wastewater. From here the vertical integration of the service: one single company has to be in charge of the whole water cycle that also includes the sewerages and the treatment, to assemble responsibility and ability of planning and programming. The main objectives of law 36/94 are:

- unification of the management of the service;
- separation of the functions of management;
- planning of the levels of service;
- IWS management following efficiency criteria;
- coverage costs through the tariff.

More specifically it foresees the reorganization of the water services through the constitution of Optimal Territorial Basins (OTB) defined by the Regions and individualized through the territorial and functional integration of the different activities of the integrated cycle (Massarutto [3]). Based on the necessity to industrialize the sector and, at the same time, on the impossibility of avoiding a natural monopoly, the law imposes at least the exploitation of all possible economies of scale and scope, at reaching a competition for the market as a surrogate of the impossible competition in the market. For this reason, a plan must be defined analyzing the pre-existing management, that constitutes however the point of departure of the plan of future management. After the recognition, the plan determinates the targets' project or the necessary interventions to bring the service to that levels of efficiency imposed by the Law and their temporal distribution, estimating both the operational costs and the improvements of the possible effectiveness and efficiency, regarding on the base of the future program of the infrastructural and organizing interventions. The tariff development goes on through a series of operations that allow the adaptation of two different aims: from one side the necessity to bring the level of service to a desired value of efficiency, from the other one the obligation to contain the tariff increases within preset limits with the contemporary coverage of the costs of management (Meucci and Peruzzi [4]). More specifically, the predisposition of the plan can be realized through the following phases showed in Figure 1:

Phase 1. *The recognition*

The analysis of the works and the existing infrastructures is necessary to give the service levels that the existing structures are able to assure before the reforming plan began to be implemented, the analysis of the productive capacity, the possible risk sources and precariousness, as well as indications concerning the water balance.

Phase 2. *The levels' definition of service*

After the recognition, the territorial Authority will fix the levels of service necessary for the satisfaction of the usage defining the characteristics of the desired service and its quality parameters. The comparison between the existing situation and the desired one allows consideration of the critical situations on which it is necessary to intervene with the plan of the investments. It is now possible to give a dimension and a priority to the problems and therefore to define the purpose of every intervention in terms of quantitative and qualitative objectives (Fucci et al. [5]).

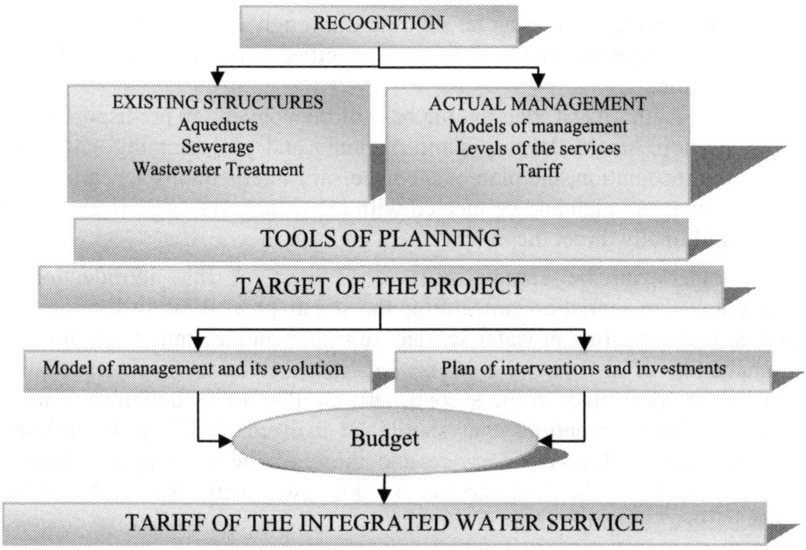

Figure 1: Principal phases for the predisposition of a territorial basin plan.

Phase 3. *The planning of the interventions*

The Plan is now able to establish the real program of the interventions and to identify the plan of the investments; these must be realized to fill the difference among the desired levels of service and those that the existing structures can reasonably assure in the period of management. The identification of every intervention is connected to the achievement of specific objective, so that we can evaluate the cost of each operation.

Phase 4. *The plan and the tariff development*

After the recognition and determination of the project's targets, the analysis of the pre-existing management is required. Considering the operational costs of the first year of the integrated management and the possible improvements of effectiveness and efficiency in the following years, it is possible to identify the operational management costs in the thirty years of the Plan. The aim of the plan is to raise the level of water services, over the period specified by the plan, to that set by the Basin Authority. The tariff development is divided in a series of operations that allows the adaptation of two different purposes: at one side the necessity to bring the level of service to a datum value of efficiency, from the other the obligation to contain the tariff increases within preset limits with the contemporary coverage of the costs of management.

There are three fundamental actors of the institutional and industrial reorganization of the Integrated Water System: the Region, the Authority of Plan and the Service Administrator (a single enterprise). Galli's Law provides that the Plan has to assign the management of the Integrated Water Service through a series of preliminary operations:

- ✓ The management must be submitted through a convention that answers to the type scheme and to well specified criteria predisposed by the Region.
- ✓ The Authority of Plan, on the base of the convention predisposed by the Region, has to develop some planning and programming activities as the recognition, the plan of the necessary interventions accompanied by a financial plan and connected with the managerial organization model, and finally direct the management control.

This paper can be summarized as follows: in this introduction the fundamental characteristics concerning the depth process of institutional and industrial reorganization of water service, sewerage and treatment determined by the emanation of the law 5/1/94 n. 36, "Dispositions in subject of water resources" are presented; in the second part, the Plan of an Italian case and the program of the interventions that should be realized to filling the difference among the desired levels of service and those of the existing structures are illustrated. Finally in the third part the evolution of a tariff capable of sustaining the execution of the program is presented.

2 An Italian OTB plan of investments

Abruzzo Italian Region is divided into 6 OTBs: Pescara, Marsica, Teramo, Peligno Alto Sangro, Chieti and L'Aquila. OTB 1 (L'Aquila) is constituted by 37 Municipalities, it has a surface of 1856 Km^2, with a resident population of 103995 inhabitants for a density of 56 ab/Km^2. Considering the program of interventions that it will realized with precise priorities we can defined the plan of investments that will fill the difference among the desired levels of service and those that the existing structures can assure in the period of management. Such plan with a life of 30 years, foresees investments for a total of 151 millions of Euro.

Figure 2 shows the chronology of the total investments that are more consistent in the first 5 years. It is possible to consider investments for typology of works: external aqueducts, distribution networks, sewerage, treatment and various expenses. The investments in the external aqueducts (including catchment, liftings, telecontrol, ordinary and extraordinary maintenance) are more raised in the first 4 years for the primary necessity to realize new pipes. As it regards the investments in distribution (including counters, network enlargement, tanks, ordinary and extraordinary maintenance), it is preferred to spread the expenses for the construction of a new tank (10.000 m^3) between 2015 and 2019 not to already increase the huge investment in the first years of plan. The investments in sewerage (network enlargement and extraordinary maintenance) are enough constant in the time. Crucial points of the economic technical plan are the treatment investments defined in the previous Italian plan (Transitional Plans); it has been necessary to compress these investments between 2002 and 2005 (5,5 millions of Euro per year). It can be noticed that the other investments are particularly heavy for the construction of the new head office administrator and for the modernization of the machineries.

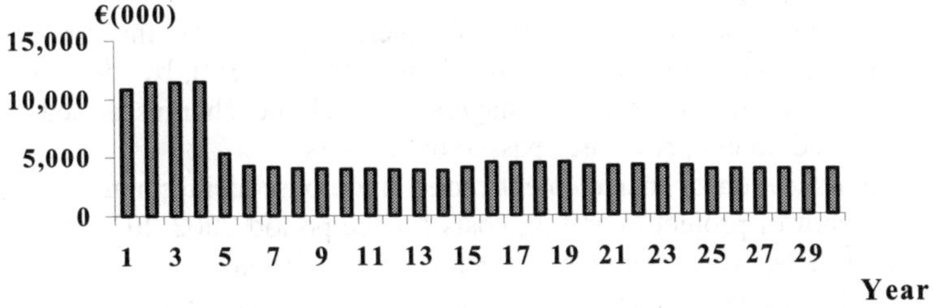

Figure 2: The plan investments.

3 The development of the tariff Plan

Once the investments and their temporal distribution have been defined and estimated, the Normalized Method adopted by the Ministry of Public Works with D.M. of 1/08/96 allows the determination of the integrated water service tariff introducing a mechanism of temporal adjustment based on the price-cap method (Passatelli [6], Peruzzi [7], Christopher [8]). This system is based on the principle that the annual tariff increase must be arranged by the administrator and the public authority on the base of the plan of investments (Baldini [9]). The maximum annual increase of the tariff depends to the rate of inflation and also to a parameter that keeps in mind the managerial objectives measured in terms of recoveries of efficiency and attainment of the service standards. The tariff in the current year is given by (Bonanni et al. [10]):

$$T_n = T_{n-1}\left(1+\pi+K\right) \qquad n=1,\ \dots,\ T; \tag{1}$$

where T_{n-1} is the tariff in year $n-1$ function of C, A and R where C represents the component of the real operational costs, A is the component of the costs of depreciation and R it is the component related to the return of the invested capital, π is the annual officially planned inflation rate and K is the "limit of price". The estimated operational costs of the plan will be compared with the operational reference costs to be able to foresee well determined recoveries of efficiency. The model allows calculation of the tariff and to assure the integral coverage of the investment and management costs estimated in the plan, compared with those of reference. Being (1) a cyclical model, it is necessary to define the point initial T_0 or rather the pre-existing middle tariff. This is given by the ratio between all the revenues of the pre-existing managements, comprehensive of the canons related to wastewaters and the water sold. The Normalized Method foresees that the real average tariff has to immediately make reference to the annual exercise before to the adoption of the financial economic plan of the Integrated Water Service, in exercise from 2002. This method basically requires the following steps (Supervisory Committee on the Use of Water Resources [11]):

- determination of the weighted average tariff of the pre-existing managements (indicated by the abbreviation TMPP), by identifying the invoices sales of these managements and the charges added by the Method unless previously passed on to users;
- identification of management model and consequent quantifying of the amount of project operating costs for the period 2002-2031 (personnel, materials, water purchased electric energy, rentals, leasing, etc.) to be compared with those modelled on the basis of two aims approved by the Method: determination of percent cost reduction to enhance efficiency and the evaluation of the congruence between the plan forecasts and the results of econometric models of the cost of water supply, sewage and treatment;
- identification of annual expenditure an investments, and consequent determination of the component to be charged to depreciation and return on invested capital (a rate of 7% is applied);
- after comparing the operating costs, depreciation and return on invested capital, the real average tariff is then determined by dividing the three components by the estimated volume supplied;
- the resulting price increase must be contained within a maximum admissible value; should be not the case, the initially planned interventions are remodulated until all the percent tariff increase planned in the project are lower than the maximum allowed ceiling

For the determination of the admissible maximum value of the index percentage of the limit of price (*K*) the Method (art. 5 of the DM 1/08/96) gives a table of decreasing values in comparison to the value of the tariff related to the preceding year. Such verification is to safeguard the use from increases too elevated tariffs. Then it is necessary to submit the determined tariff development to the last verification required by the Method that it consists in the respect of the "limit of price" established. It proceeds determining, for every year, the relationship between the real tariff of the considered year and that of the preceding year and it is compared with the limit of price, available in the art. 5 of the Method and related to the considered year. The verification has put in evidence the overcoming of the limit K in different years of the plan, particularly in that years characterized by elevated investments in the treatment sector. The temporal development of the investments and that of the OTB tariff are shown in Figure 3. In the case in examination, therefore, the tariff increases are not contained within the maximum limits of price allowed and this doesn't allow the application of the individualized strategic plan. Such plan foresees investments for a total of over 151 million € in thirty years particularly strong in the first four years; this plan is too strong if compared with the dynamics of the consequential tariff entrances. This is due to the following reasons synthetically exposed. First of all it is necessary to notice that the considered OTB has a population density very low; this determines elevated networks lengths in the integrated water service (some municipalities are geographically situated to elevated distance from the Provincial capital and have a population that doesn't often overcome the 200 unities). Besides, part of the distribution networks, report levels of insufficient functionality due to the networks age and of their insufficient state of maintenance; it appears clear, therefore, that is necessary to provide, where possible, to an immediate substitution or maintenance. Finally it observed that the plan, of which to the article 141 of the Law 388/2000 (Financial Law for 2001), foresees strong investments in the depurative sector and in the sewerage system especially in the first four years of the management not completely covered by the tariff development.

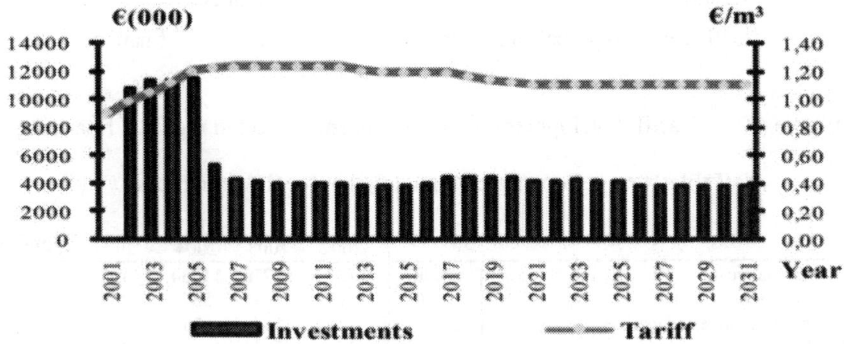

Figure 3: Not user sustainable tariff development.

To allow the verification of the strategic economic plan a regional financing has been proposed in order to cover a substantial part of the investment in the treatment sector. Such financing, around 21 million of euros, allows the not overcoming of the limit of price K and, in such way, to make the tariff growth sustainable for the customers.

The new comparison between real tariff growth and proposed investment plans assumes the evolution described in Figure 4. Thanks to the regional financing the temporal development of the investments and that of the tariff growth, satisfy both the demands of improvement of the offered service and the necessity to remunerate adequately the activity of the future administrator. Besides, in a period of thirty-year, levels of service are qualitatively and quantitatively reached to resolve the actual criticism of the studied OTB. In Table 1 the fundamental economic characteristics of the Plan for L'Aquila OTB are shown. For brevity the data are related to the first five years of the Plan and those following on a range of five years base.

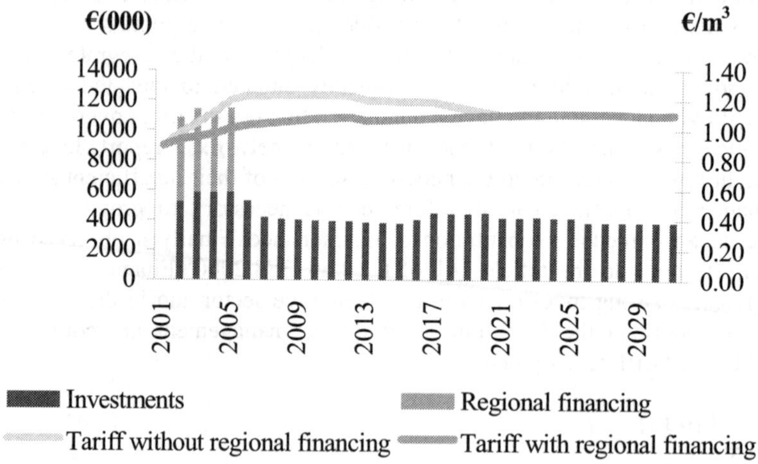

Figure 4: Tariff development, investments and expected regional financing.

Table 1: Economic-financial plan for the Italian case

		2002	2003	2004	2005	2006
Operational costs	€(000)	11973	11710	11534	11536	11362
Amortizations	€(000)	179	368	557	752	1.013
Return of the invested capital	€(000)	178	547	923	1.287	1.616
Total	€(000)	12329	12624	13014	13575	13990
Supplied volume (forecasted)	m^3(000)	13118	13227	13336	13445	13554
Real Tariff	€/m^3	0,94	0,95	0,98	1,01	1,03
Annual Tariff variation (K)	%	4,89%	1,55 %	2,25%	3,46%	2,23%
Maximum annual admitted variation (K)	%	7,5%	5,0%	5,0%	5,0%	5,0%
Starting Tariff (TMPP)	0,89					

		2011	2016	2021	2026	2031
Operational costs	€(000)	10912	10322	10092	9865	9643
Amortizations	€(000)	1.705	2.109	2.632	3.160	3.635
Return of the invested capital	€(000)	2.586	3.275	3.979	4.413	4.571
Total	€(000)	15203	15706	16703	17438	17849
Supplied volume (forecasted)	m^3(000)	14099	14643	15188	15733	16169
Real Tariff	€/m^3	1,08	1,07	1,10	1,11	1,10
Annual Tariff variation (K)	%	0,33%	0,25 %	0,36%	-0,01%	0,37%
Maximum annual admitted variation (K)	%	5,0%	5,0%	5,0%	5,0%	5,0%

With reference to tariff evolution during the plan period, it should be noted that there is a further confirmation of the initially growing and later stable trend of the intertemporal tariff profile due to the combination of two factors: on the one hand, the concentration of interventions in the first four years after the assignment of the integrated water service and on the other hand the pricing mechanism which, for two components out of three, is linked to the net accounting value computed for the completed works.

4 Conclusions

In this paper a model of management of the Integrated Water Service has been introduced. Such model represents one of the pilot projects in Italy. In fact, although the actual legislation is complete, the managerial model is still incomplete (Muraro [12] and Marra [13]). This plan represents therefore a pilot project that can be followed by the other Optimal Territorial Basins. To conclude our paper a geographical analysis can be performed to illustrate the results of tariff mechanism in 41 Basin Plans in Italy (Figure 5). Compared with absolute tariff levels it is observed that the values in the South and in the Islands are on average higher than in the other basins of the country (Supervisory Committee on the Use of Water Resources, [11]). The possible reasons for this difference may be the different forecast project operating costs which, in some situations, represent the only tariff based charges. Our contacts with the regional authorities evidenced many subjective elements, which confirm that the willingness to implement the presented reform is widespread, even if not unanimous. Finally, the last governments have explicitly taken a stand in favour of the Integrated Water Service reform and have promised to accelerate it by studying and introducing some relevant procedural simplifications. Moreover the effort to increase the use of project financing in Italy looks promising, also in the water sector. At the same time, some concern must be expressed for several particular initiatives taken by municipalities and other institutions, which aim, by using the project financing, at building new infrastructures for the water service without following the procedures set out by Law 36/94: such initiatives may indeed postpone or even hinder the real implementation of the reform.

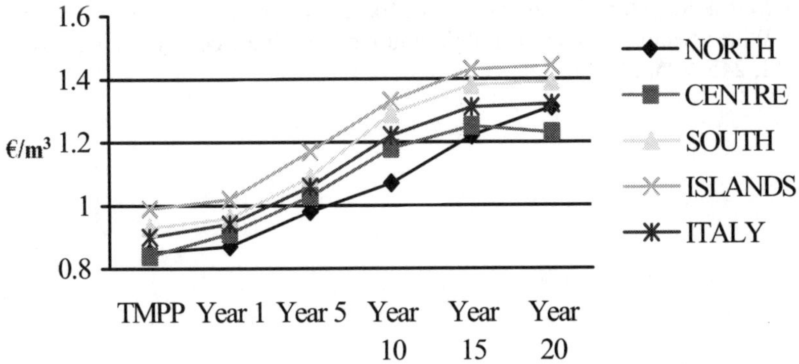

Figure 5: Tariff development per reference geographical area.

References

[1] Barbese, E. and Meucci, F., Note sulla salvaguardia, coordinamento e integrazione di funzioni (ex art. 9, comma 4, legge 36/94), Papers Proaqua, Report Number 8, 1997.
[2] Caselli, R. and Peruzzi, P., Servizi idrici fra regolazione e mercato, Papers Proaqua, Report Number 19, 1998.
[3] Massarutto, A., Il ciclo integrato delle acque. Regole di mercato e modelli operativi a confronto, Franco Angeli, 2001.
[4] Meucci, F. and Peruzzi, P., Manuale del Piano di Ambito per il Servizio Idrico Integrato, Franco Angeli, 1998.
[5] Fucci, G., Lubello, C. and Meucci, F., Livelli di servizio - Descrizione, analisi e commento dei livelli di servizio per i Piani delle Autorità di Ambito, Papers Proaqua, Report Number 15, 1998.
[6] Passatelli, M., Struttura della regolamentazione tariffaria in applicazione della legge 36/94, Report Number 1, 1995.
[7] Peruzzi, P., Le tariffe dei servizi idrici, Papers Proaqua, Report Number 7, 1996.
[8] Christopher, M. G., Logistics and Supply Chain Management; strategies for reducing costs and improving services, London, Pitman Publishing, 1992.
[9] Baldini, D., Il finanziamento degli investimenti nel settore idrico, Papers Proaqua, Report Number 37, 2001.
[10] Bonanni, A., Gastaldi, M. and Rocca, C., Riorganizzazione e gestione del Servizio Idrico Integrato, Franco Angeli Editore, 2003.
[11] Supervisory Committee on the Use of Water Resources, Annual Report to Parliament on the State of the Water Service, 2003.
[12] Muraro, G., Water services and water policy in Italy, in Sustainable Development and Environmental Management Experiences and Case Studies (Clini, Musu and Gullino Eds.), Springer Netherlands, 65–90, 2008.
[13] Marra A., Internal Regulation by Mixed Enterprises: the case of the Italian Water Sector, Annals of Public and Cooperative Economics, Vol. 78, No. 2, 245–275, 2007.

Reducing diffuse water pollution by tailoring incentives to region specific requirements: empirical study for the Burdekin River basin (Australia)

R. Greiner & O. Miller
River Consulting, Townsville, Queensland, Australia

Abstract

Australia is facing many environmental problems caused by agricultural diffuse pollution. Policies and programmes are being developed for landholders to improve environmental performance. One tool for achieving environmental improvements is the design and promotion of 'best management practices' (BMPs). These are conservation practices aimed at reducing diffuse pollution from agricultural lands and thus improving end-of-catchment water quality. A suite of grazing region-specific BMPs were developed for the Burdekin Dry Tropics region in north-eastern Australia. While they were developed in a consultative fashion, there was no explicit consideration of knowledge of adoption processes or supporting incentives. This paper utilises the data from an earlier grazier survey to explore what extent landholder motivations influence the adoption of BMPs and to gauge landholder preferences for incentives. The results highlight critical correlations between landholder goals, barriers to adoption of conservation practices, and preferred incentives to help overcome barriers. We conclude that a sound understanding is required not only of regional environmental issues but also the people and business situations which control diffuse pollution so as to tailor programmes aimed at improving regional environmental performance.

Keywords: diffuse pollution, non-point pollution, water quality, conservation practices, incentives, adoption, grazing, Burdekin River basin, Great Barrier Reef, empirical research, correlations, factor analysis.

1 Introduction

Agricultural land use generates or contributes to a number of environmental problems, such as water quality decline, by emitting pollutants into the natural environment. Agriculture-related processes causing water quality decline include soil erosion, and runoff and percolation of plant nutrients and pesticides. The difficulty in improving water quality is two-fold. Farms are small portions of landscapes embedded in large-scale biophysical and ecological processes at regional scale and therefore farming activities are commonly associated with externalities [1]. This means the resulting costs are either borne by society or downstream water users and therefore not considered by the polluter. Secondly, this type of pollution is diffuse, or of non-point source character, and while individual contributions can be minor their collective impact is often significant and accumulates over time [2, 3].

Pollution control policy is largely concerned with the design and performance of emissions-based instruments in situations where pollution sources identified and emissions can be measured accurately [4, 5, 6]. In comparison, there have been limited policy responses to diffuse (or non-point) pollution problems because of two inherent features [6, 7]. Firstly, there is a high degree of uncertainty about non-point emissions and current monitoring technology cannot attribute nonpoint emissions to particular emitters with reasonable accuracy and/or at reasonable cost. Secondly, the spatial variation in emission affects feasibility, effectiveness and cost of technical options for reducing emissions. Consequently, the policy focus has been on design-based or indirect instruments, whereby the land use activities are targeted that are thought to cause the pollution [8, 9, 10]. Indirect instruments are considered more likely to provide cost-effective control, particularly when monitoring and enforcement costs are considered.

The policy response to diffuse pollution problems must focus on the actual characteristics of the environmental problem and its spatial variability, and the human factors that constrain improvements [6]. The literature conclusively suggests the adoption of innovations, including conservation practices, by farmers is primarily driven by the extent to which the proposed practices are seen to support their goals, and the perceived riskiness of the innovations e.g. [10].

The objective of this paper is to provide, through an exploratory and descriptive case study, empirical insights into what incentives—and combinations thereof—may be best suited to generating water quality improvements in a river basin located in north-eastern Australia. It provides a contribution to the body of empirical literature as well as helping to support the design of effective and efficient regional programmes and initiatives for the Burdekin River basin.

2 Background

The Great Barrier Reef (GBR) is the largest coral reef ecosystem in the world, covering an area of 347,800 km^2 and measuring over 2000 km in length along

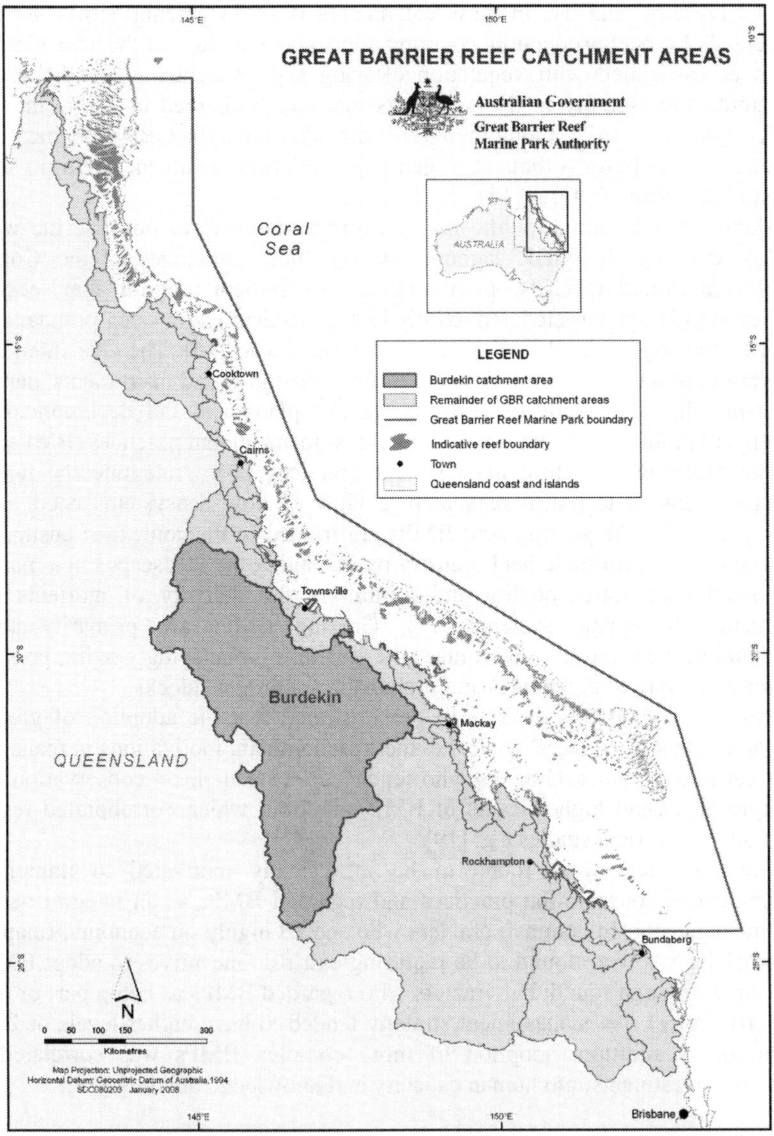

Figure 1: Burdekin River basin in the context of the GBR catchment.

the north-eastern Australian coast, fig. 1. It was designated a World Heritage Area in 1981 in recognition of its outstanding intrinsic values [11]. Its annual contribution to the economy was estimated to be AUD6.1 billion [12]. The health of the Great Barrier Reef ecosystem is critically influenced by the nutrients, sediments and other pollutants discharged into the GBR lagoon from a large number of adjacent river catchments.

The primary land use in these catchments is cattle grazing across the vast rangelands for beef production, covering approximately 98% of the area. Grazing practices associated with vegetation clearing and overgrazing are thought to contribute substantially to the sediments that are discharged by rivers into the GBR lagoon [13, 14]. Among the river basins that form the GBR catchment, the Burdekin River basin is thought to generate the largest sediment due to its large size and landscape features [15], fig. 1.

There are substantive public policy efforts underway to improve the water quality entering the GBR lagoon. Among these initiatives is the Coastal Catchment Initiative (CCI), an Australian Government program that seeks to deliver significant targeted reductions in the discharge of water pollutants to agreed 'hotspots', including the Burdekin River basin [16]. The CCI intends to deliver improvements in water quality through design-based instruments, namely assistance for catchment-based water quality plans and the development of region and industry-specific water quality best management practices (BMPs).

The philosophy of best management practices is to 'integrate the human dimension into a technical or scientific view of how ecosystems need to be managed' [17]. The grazing land BMPs are framed in the context of ensuring a sustainable and profitable beef industry by managing the landscapes in a manner that maximises water quality and minimises the delivery of nutrients and sediments to aquatic systems [17]. Grazing BMPs are primarily about maintaining grass cover to minimise soil erosion by managing grazing pressure via infrastructure (e.g. fencing) and systematic spelling paddocks.

Research by Greiner et al. [19] demonstrated that the adoption of grazing BMPs was linked graziers' goals, i.e. their predominant motivations in managing their cattle operations. Graziers who tended to score high on conservation and lifestyle goals had higher levels of BMP adoption, which corroborated results from other empirical studies e.g. [19].

Those graziers were found to be intrinsically motivated to implement recommended conservation practices and regarded BMPs as an integral part to risk management. In contrast, graziers who scored highly on economic/financial and social goals were found to be requiring extrinsic incentives to adopt BMPs. The same research found that graziers who regarded BMPs as being part of their property (price) risk management strategy tended to have higher levels of BMP adoption. In addition, adoption of more complex BMPs was correlated to graziers' investments into human capacity and knowledge building [18].

3 Materials and methods

The data used for this research originated from a survey of landholders in the Burdekin River catchment, which was conducted in late 2006 to explore social and economic dimensions of the implementation of BMPs within a regional and industry context [20].

The data set offered the opportunity for additional quantitative research into how landholder motivations and related to the preferred incentives for adoption of environmental management practices. This research focuses on the subsample of 94 grazier respondents (from a total sample of 114), which is the same subsample explored by Greiner et al. [19]. The research is of an exploratory nature due to the small subsample (<100). Average size of respondents' grazing properties was 34 506 hectares, maximum size was almost 230 000 hectares.

Basic statistics and multivariate techniques were employed for data analysis and were conducted in STATISTICA (v7.1) [21]. Graphics and tables were generated in Microsoft Excel. The alpha level for testing for statistical significance was set at 0.05 unless stated otherwise.

Principal Component Analysis (PCA) was used to (1) explore a set of variables with a view to identifying the underlying structure and (2) to simplify a large set of variables into a smaller, simpler set of factors for further analysis [22] and has been used in previous studies for similar data and outcome [23, 24]. Factor solutions with different numbers of factors were examined before structures were defined to have the most representative and parsimonious set of factors [25].

Correlation analysis (Pearson's R) was performed to test relationships between variables, which were the factors generated by PCA. Items measured using Likert-type scales were treated as continuous variables and standard parametric statistical procedures were employed [25, 26, 27]. Missing data were dealt with in a pair-wise fashion to maximise the sample.

4 Results

Survey respondents were asked to rate how strongly various factors served as impediments, preventing them from implementing more conservation practices. There were 31 impediment items, including financial, skills and knowledge, operational, personal and other constraints. The Likert-type rating scale ranged from 1='not a constraint' to 10='fundamental constraint'. The highest rating impediment items were 'initial investment cost', 'too much red tape' and 'not enough time', fig. 2, followed by 'variable climate and drought', 'not enough staff or labour', 'high ongoing costs', 'low return on investment', and 'lack of government incentives'. This suggests that predominantly financial considerations, but also institutional, risk and operational aspects act to constrain the implementation of BMPs.

PCA was employed to reduce the item set to seven impediment factors, which explained 65% of variance. The factors included: Lack of external support and incentives for implementation; practices not recommended by industry; risk and uncertainty; ill-fit with current operating practices; not enough staff/labour and time; and high cost of implementation.

Correlation analysis between the impediment factors and graziers' goals revealed no relationship between the level of lifestyle/conservation motivation and any impediment factors, tab. 1. In contrast, graziers with higher social motivations tended to be constrained by a lack of support from government/society/industry, perceived lack of fit of BMPs with the current operating system and direct costs associated with implementation. The latter constraints were also prevalent among highly economically/financially motivated

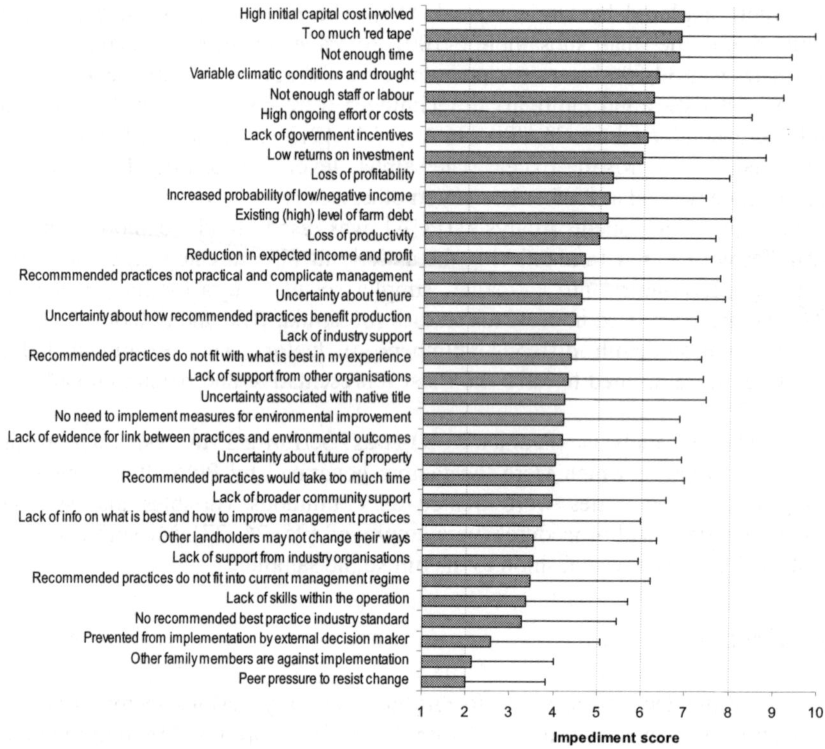

Figure 2: Impediments to the implementation of conservation practices.

Table 1: Correlation matrix between grazier goals and impediment factors. 69 <= N <= 81.

	Grazier goals		
Impediment factors	Conservation & lifestyle	Economic / financial	Social
Reduced farm productivity and profitability	-.0190	.1185	.1619
Lack of support by government & industry	.1145	.2270*	**.2951****
BMPs not recommended by peak bodies	.0869	.2298*	.1168
Risk and uncertainty (climate/markets/etc)	-.0120	.2279*	.1299
Ill-fit with current operating system	-.0242	**.4709*****	**.2727****
Not enough time or staff/labour	-.0947	.2024*	.1404
Direct costs of implementing/doing practices	.1764	**.4594*****	**.2835****

graziers. At the $p<0.1$ level these graziers also tended to perceive a series of other impediments.

Correlation analysis showed that the more conservation and lifestyle motivated graziers tended to be the ones with lower equity, while no relationship between farm debt or equity and other grazier goals could be established.

Survey respondents were asked to rate how effective they perceived each of a suite of incentive instruments to be in addressing their impediments to the adoption of conservation practices. The list included 28 incentive instruments from various categories including financial incentives; extension, education and research; regulation; voluntary and industry measures, and recognition incentives. The rating scale ranged from 1='completely ineffective' to 5='highly effective'. The mean ratings and standard deviations are shown in fig. 3.

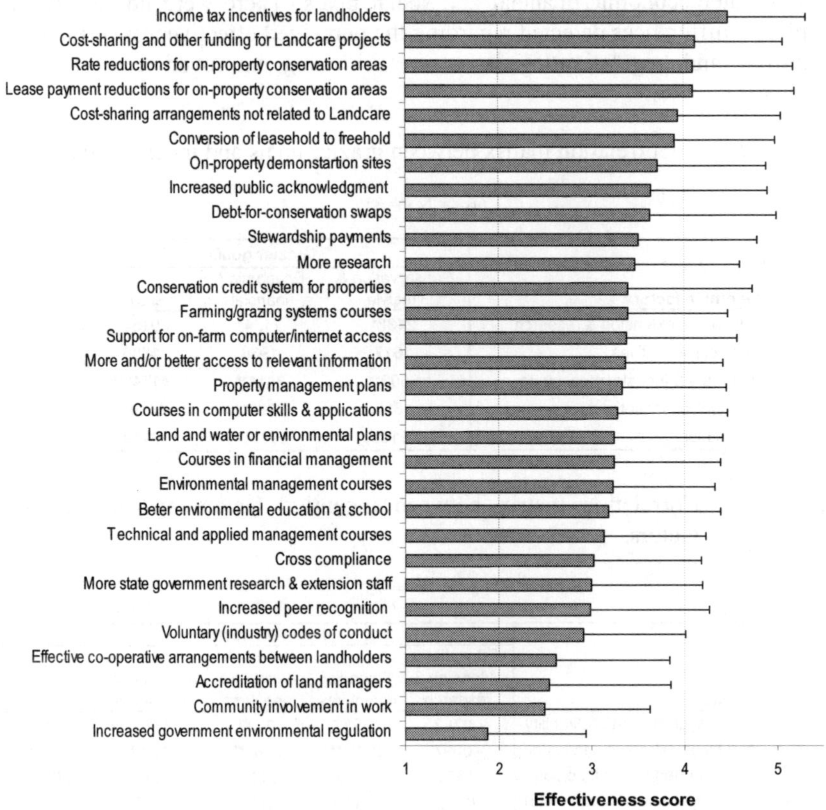

Figure 3: Perceived effectiveness of incentive instruments.

Graziers regarded income tax incentives as the single most effective incentive, followed by a suite of other financial incentives. The prominence given to financial incentives and the overarching desire for tax incentives is consistent with results from an earlier survey by Greiner et al. [29], who noted the strong influence of business advisors and tax agents on farm management decisions across the Burdekin River basin. However, increased security of tenure (conversion of leasehold to freehold), increased public acknowledgement of conservation efforts by landholders, on-property demonstration sites, and more research also received moderate to high effectiveness scores.

PCA was performed on the incentive items and produced five incentive factors, which explained >59% of the variance. The composition of the factors was virtually identical to the conceptual grouping of incentives adopted for the survey [20].

Correlation analysis was conducted between grazier goals and incentive factors, tab. 2. Voluntary and/or industry measures tended to be rated as more effective by more highly motivated graziers, irrespective of the types of goals they pursued. Recognition incentives tended to be rated more highly by graziers with stronger economic/financial and social goals. There were no statistically significant differences detected regarding the perceived effectiveness of financial incentives and regulation/legislation in achieving improved environmental outcomes.

Table 2: Correlation matrix between grazier goals and incentive factors.

66 <= N <= 81.

	Grazier goals		
Incentive factors	Conservation & lifestyle	Economic / financial	Social
Education, extension & research	.2138*	.2124*	.0154
Financial incentives	.1617	.1938	.1885
Voluntary & industry meaasures	**.2602****	**.2414****	**.4174*****
Recognition incentives	.1653	**.3069****	**.2430****
Regulation	-.0160	.0532	.0419

Table 3: Correlation matrix between incentive factors and impediment factors.

65 <= N <= 75.

	Incentive factors				
Impediment factors	Education, extension & research	Financial incentives	Voluntary & industry measures	Recognition incentives	Regulation
Reduced farm productivity and profitability	-.0783	-.0987	.0699	.0508	-.0245
Lack of support by government & industry	-.0097	**.2471****	**.2569****	.1802	**.2537****
BMPs not recommended by peak bodies	.1624	.1403	**.3223*****	.0246	.2055*
Risk and uncertainty (climate/markets/etc)	.0786	-.0119	**.4475*****	.1577	-.1173
Ill-fit with current operating system	.0357	.0852	**.2718****	**.3353*****	-.0446
Not enough time or staff/labour	-.0813	-.0205	.0110	.0292	-.0380
Direct costs of implementing/doing practice	.1551	.1849	**.2746****	**.3222*****	.0505

Correlation analysis between incentive factors and impediment factors shows that those respondents who tended to rate voluntary and industry measures as highly effective, tended to be constrained by a number of impediments—with the exception of farm profitability and not enough time/staff/labour, tab. 3. Those who tended to rate 'lack of support' more highly as an impediment rated financial incentives, voluntary measures as well as regulation/legislation as being more effective. Recognition incentives were rated highly by those who regarded ill-fit with current operating system and direct cost has major impediments.

5 Discussion and conclusions

The approach, in Australia, to address diffuse water pollution problems in key regions, such as the GBR catchment, by developing region and industry-specific conservation practises in consultation with landholders provides a first step to addressing the discharge of water pollutants from agricultural lands. The approach is consistent with recommended policy approaches to diffuse pollution problems. However, in itself this does not provide a sufficient strategy, as results from this exploratory case study for graziers in the Burdekin River catchment indicate. Rather, a strategic mix of incentives is required to facilitate the broad-scale adoption of conservation practices, at the critical extent and rate of change required for discernable improvements in the water quality delivered by the Burdekin River to the GBR lagoon in order to preserve the intrinsic values that underpin its World Heritage status [29].

BMPs have been adopted particularly by those graziers who pursue lifestyle and conservation goals and are intrinsically motivated to adopt conservation practices. However, graziers with high economic/financial and social motivation require external incentives, specifically where changes to the current operating system are comprehensive and risky [30]. They require incentives because they see themselves constrained by particularly financial impediments and perceive a disconnection between conservation practices and industry expectations. Conservation practices need to be endorsed or recommended by industry bodies to penetrate the mainstream grazing community.

The results also suggest that there is quite rightly a focus on financial incentives, which are favoured by all landholders irrespective of their main goals and constraints. While income tax incentives provide a blanket and substantive financial lure, they need to be complemented by region-specific cost sharing programmes designed to overcome the financial barriers of the initial investment and also act as a risk premium. There is already a focus on developing small-scale regional cost sharing programmes with landholders to foster environmental performance across various environmental domains. However, the design of conservation practices as well as financial programmes will be more effective and efficient if they, too, are linked more strongly with and supported by regional industry organisations.

Incentives do not always have to take the shape of financial incentives. An often ignored motivation is recognition. Recognition, by peers and the community, of conservation efforts serves as a powerful incentive for highly motivated graziers. The reverse may also be true, i.e. that a lack of recognition may lead to a decline of the voluntary (and 'free') provision of environmental services by landholders.

Extension, education and research play a critical supporting role in the adoption process. New paradigms and social imperatives require different ways of managing the land, which graziers need to comprehend and be skilled to implement and maintain, while trying to steer their often very large properties through volatile climatic and market conditions. Finally, appropriate regulatory intervention is also a key component of an integrated incentives strategy to combat diffuse source water pollution.

Acknowledgements

The initial survey upon which this paper is based was funded by the Coastal Catchment Initiative and the Burdekin Dry Tropics NRM through funding provided by the National Action Plan for Salinity and Water Quality. The map was provided courtesy of the Spatial Data Centre, Great Barrier Reef Marine Park Authority.

References

[1] Vatn, A. and Bromley, D.W., Externalities—a market model failure. *Environmental and Resource Economics*, **9**, pp. 135–141, 1997.
[2] Ferrier, R.C., D'Arcy, B.J., MacDonald, J. and Aitken, M., Diffuse pollution – what is the nature of the problem? *Journal of the Chartered Institution of Water and Environmental Management*, **19(4)**, pp. 361–366, 2005.
[3] Taylor, M.L., Adams, R.M. and Miller, S.F., Farm-level response to agricultural effluent control strategies: the case of the Willamette valley. *Journal of Agricultural and Resource Economics*, **17(1)**, pp. 173–185, 1992.
[4] Shortle, J.S. and Horan, R.D. The economics of nonpoint pollution control. *Journal of economic surveys*, **15(3)**, pp. 255–289, 2001.
[5] Russell, C.S. and Powell, P.T., Practical considerations and comparison of instruments of environmental policy. *Handbook of environmental and resource economics*, ed. J. van den Bergh, Edward Elgar Publishing Inc: Cheltenham UK, 1999.
[6] Falconer, K.E., Managing diffuse environmental contamination from agricultural pesticides: an economic perspective on issues and policy options, with particular reference to Europe. *Agriculture, Ecosystems and Environment*, **69**, pp. 37–54, 1998.
[7] Shortle, J.S., Horan, R.D. and Abler, D.G., Research issues in nonpoint pollution control. *Environmental and Resource Economics*, **11(3-4)**, 571–585, 1998.
[8] Helfand, G.E. and House, B.W., Regulating nonpoint source pollution under heterogeneous conditions. *American journal of Agricultural Economics*, **77**, pp. 1024–1032, 1995.
[9] Ribaudo, M. O., Heimlich, R., Claassen, R. and Peters, M., Least-cost management of nonpoint source pollution: source reduction versus interception strategies for controlling nitrogen loss in the Mississippi Basin. *Ecological Economics*, **37**, 183–197, 2001.
[10] Pannell, D.J., Marshall, G.R., Barr, N., Curtis, A., Vanclay, F., and Wilkinson, R., Understanding and promoting adoption of conservation

technologies by rural landholders. *Australian Journal of Experimental Agriculture*, **46**, 1407–1424, 2006.

[11] Wu, J. and Babcock, B.A., Spatial heterogeneity and the choice of instruments to control nonpoint pollution. *Environmental and resource economics*, **18**, pp. 173–192, 2001.

[12] Lucas, P.H.C., Webb, T., Valentine, P.S. and Marsh, H., *The outstanding universal value of the Great Barrier Reef World Heritage Area.* Great Barrier Reef Marine Park Authority, 1997.

[13] Access Economics, *Measuring the economic and financial value of the Great Barrier Reef Marine Park.* Great Barrier Reef Marine Park Authority: Canberra, 2005.

[14] Brodie, J.E., McKergow, L.A., Prosser, I.P., Fumas, M., Hughes, A.O. and Hunter, H., *Sources of sediment and nutrient exports to the Great Barrier Reef World Heritage Area* (*ACTFR Report No. 03/11*). Australian Centre for Tropical Freshwater Research, James Cook University: Townsville, 2003.

[15] Fumas, M., *Catchments and Corals: Terrestrial Runoff to the Great Barrier Reef.* Australian Institute of Marine Science and CRC Reef Research Centre: Townsville, 2003.

[16] Roth, C.H., Lawson, G. and Cavanagh, D., *Overview of key natural resource management issues in the Burdekin catchment: with particular reference to water quality and salinity (Burdekin condition study phase 1).* Burdekin Dry Tropics Board: Townsville, 2002.

[17] The Australian Government, http://www.environment.gov.au/coasts/pollution/cci/index.html

[18] Coughlin, T., O'Reagain, P. and Nelson, B., *Draft review of current and proposed grazing land best management practices for achieving water quality objectives in the Burdekin catchment (Grazing land management for Burdekin water quality outcomes).* ACTFR, James Cook University: Townsville, 2006.

[19] Greiner, R., Miller, O. and Patterson, L., The role of grazier motivation and risk attitudes in the adoption of grazing best management practices. *Proc. of the 52nd Annual Conference of the Australian Agricultural and Resource Economics Society*, 2008.

[20] Chouinard, H.H., Paterson, T., Wandschneider, P.R., and Ohler, A.M., *Will Farmers Trade Profits for Stewardship? Heterogeneous Motivations for Farm Practice Selection.* School of Economic Sciences, Washington State University, 2006.

[21] Greiner, R., Lankester, A. and Patterson, L., *Incentives to enhance the adoption of Best Management Practices by landholders: Achieving water quality improvements in the Burdekin River Catchment (Report to the Burdekin Dry Tropics NRM and Coastal Catchment).* River Consulting: Townsville, 2007.

[22] Statsoft, *STATISTICA System Reference*. StatSoft: Tulsa, 2001.

[23] Diekhoog, G., *Statistics for the Social and behavioural Sciences: Univariate, Bivariate, Multivariate,* Wm. C. Brown Publishers: IA, 1992.

[24] Flaten, O., Lien, G., Koesling, M., Valle, P.S. and Ebbesvik, M., Comparing risk perceptions and risk management in organic and conventional dairy farming: empirical results from Norway. *Livestock Production Science*, **95(1-2)**, pp. 11-25, 2005.
[25] Maybery, D., Crase, L. and Gullifer, C., Categorising farming values as economic, conservation and lifestyle. *Journal of Economic Psychology*, **26(1)**, pp. 59–72, 2005.
[26] Lien, G., Hardaker, B.J. and Flaten, O., Risk and economic sustainability of crop farming systems. *Agricultural Systems*, **94 (2)**, pp. 541–552, 2007.
[27] Meuwissen, M.P.M., Huirne, R.B.M. and Hardaker, J.B., Risk and risk management: an empirical analysis of Dutch livestock farmers. *Livestock Production Science*, **69(1)**, pp. 43-53, 2001.
[28] Patrick, G.F. and Musser, W.N., Sources of and responses to risk: factor analysis of large-scale US cornbelt farmers. *Risk management strategies in agriculture (Volume 7)*, eds. R.B.M. Huirne, J.B. Hardaker, and A.A. Dijkhuizen, Wageningen Agricultural University: Wageningen, 1997.
[29] Greiner, R., Stoeckl, N., Stokes, C., Herr, A., and Bachmaier, J., *Natural resource management in the Burdekin Dry Tropics: social and economic issues (Report for the Burdekin Dry Tropics NRM Board)* River Consulting: Townsville, 2003.
[30] Novotny, V., *Water quality (2nd edition): Diffuse pollution & watershed management,* John Wiley and Sons: New York, 2003.
[31] De Buck, A.J., van Rijn, I., Roling, N.G. and Wossink, G.A.A., Farmers' reasons for changing or not changing to more sustainable practices: an exploratory study of arable farming in the Netherlands. *The journal of Agricultural Education and Extension*, **7(3)**, pp. 153–166, 2001.

A productivity analysis of the Italian gas retail market

G. Capece[1], L. Cricelli[2], F. Di Pillo[1] & N. Levialdi[1]
[1]*Dipartimento di Ingegneria dell'Impresa, Università di Roma "Tor Vergata", Italy*
[2]*Dipartimento di Meccanica, Strutture, Ambiente e Territorio Università di Cassino, Italy*

Abstract

Owing to the process of liberalization, the European energy markets have undergone a radical transformation over the last decade.

In this paper we analyse the Italian natural gas market, focusing in particular on retail with the objective of assessing the performance of the natural gas retail operators whilst also taking into account the strategies adopted after the liberalization process.

To that aim, company performance is analysed using the Edgeworth index which enables an assessment of levels of both profitability and productivity. Two types of comparisons are carried out based on the processed data. The first of which concerns the size of the company and second regards the degree of specialisation or diversification of the services offered.

Keywords: gas market liberalization, Italian gas retail operator strategies, productivity analysis, Edgeworth index.

1 Introduction

The liberalization process of the gas market in Europe began in the late 1990s with the first gas directive (Directive 98/30/EC) [1], which established common rules regarding the transmission, storage, supply and distribution of natural gas. This process ended in 2003 with the Gas Directive 2003/55/EC [2], which is the European gas legislation in force at present.

The European Gas directive requires obligatory unbundling of transport and trading: i.e. all gas operators belonging to the member states must separate their gas transportation and trading functions into separate companies.

WIT Transactions on Ecology and the Environment, Vol 108, © 2008 WIT Press
www.witpress.com, ISSN 1743-3541 (on-line)
doi:10.2495/EEIA080051

Another key aspect of the liberalization directive is based upon the concept of 'eligible customers' which is a category of customers who have the freedom to choose their supplier and the right to acquire gas at a competitive price [3].

The European Gas directive was transposed into Italian law by Legislative Decree n.164/2000, known as the Letta Decree [4], which laid down important guidelines concerning the definition of the eligible customers, competition, and conditions of reciprocity.

The retail market has been chosen as the subject of this research paper since it is the only sector to be opened to free competition. The strategic choices made following liberalization in this sector are of much interest and thus have also been analysed in this paper. In reaction to the market being opened, the retail operators have implemented a strategy of diversification to minimize costs or have initiated mergers in order to consolidate their own position by means of integration with other firms distributed geographically across the nation. Such mergers have produced many multi-utility companies which also compete in various other energy sectors, such as the supply of water and electricity.

The purpose of this paper is to assess performance of natural gas retail operators by focusing on their productivity, while also considering the strategies adopted after the liberalization process. The analysis of performance is carried out using the Edgeworth index which permits an assessment of the level of company profitability and productivity.

After examining the literature this index was considered to be the most suitable for a retail market performance analysis as it allows comparisons of companies to be made both on the basis of the input and output prices on the one hand, and the productivity on the other. The main aim of this type of comparison is to determine whether the profits are exclusively due to a high mark-up, or rather mainly to a high level of productivity.

Starting from a previous research [5], our analysis takes two types of comparison into account in which the first considers the company size and the second deals with the degree of specialisation or diversification of the services offered. On analysing the results, the discriminatory factors in determining the performance trend were identified as being the company size, or the core business, or a combination of both of these.

This research paper is organized as follows: Section 2 provides a review of the literature relating to utility performance; Section 3 shows the data set chosen for the index application; Section 4 describes the main results of our work; Section 5 presents a comparative analysis with respect to the size of the company and the business strategy employed; Section 6 concludes.

2 The productivity analysis

The aim of this paper is to assess the performance of natural gas retail operators, focusing on three key measures of strategic performance: profit, productivity, and price differential. The methodology used expresses the three performance measures in a single unified equation utilising the Edgeworth index [6].

Alongside the surge of liberalization and regulatory reform, there has been increasing interest in the accurate evaluation of public utility performance.

Whether or not these reforms have produced the expected results may be investigated by utilising various methods of performance evaluation. Even though industry analysts have long recognized the importance of accurate performance evaluation as a cornerstone for the improvement of company performance, they have had to overcome some major difficulties regarding the evaluation of the gas industry. Although there are several distinctive performance measures which provide a range of information (such as profit, labour productivity, end-use price, etc.), these measures are not necessarily correlated. For example, higher profit may result from excessive market power rather than from higher productivity. In such cases, performance evaluation based exclusively on a single specific measure may distort the perception of the overall performance of a utility.

Extensive research has previously been conducted to analyse one specific aspect of physical performance in the natural gas market, for example efficiency or productivity. Examples of research of this nature are listed in the following: Aivazian et al. [7] and Sickles and Streitwieser [8] explore the US natural gas industry; Rushdi [9] discusses an Australian gas utility and Price and Weyman-Jones [10] study the United Kingdom's natural gas distribution sector. Unfortunately, as far as the natural gas market is concerned there has been very little research relating to the simultaneous effects of various aspects of performance, such as price and profit. Indeed, numerous studies regarding productivity have only focused on productivity performance, i.e. the ability to obtain more output with less input. Likewise, exclusively studying the price differential of a utility (comparing the output price with the input price) would only provide us with an indication of the utility's ability to maintain its profit margin and value-focused performance.

Several studies indicate that the three different measures of performance (profit, productivity and price differential) should be considered simultaneously in order to obtain an overall picture of the performance of a utility [11, 12].

In contrast to the procedures used in the above studies, the methodology utilised in this paper has two distinctive features. Firstly, using the Edgeworth index the performance difference can be calculated in terms of absolute values, which helps decision-makers understand and evaluate strategic advantages. Secondly, it allows a multilateral comparison to be made, which is particularly useful should a panel analysis be required.

In relation to the Italian market, a performance analysis of the gas distribution sector was performed by Erbetta and Fraquelli [13] who utilised the Törnqvist index on the moving average without taking all three key measures of performance into account.

Our paper follows the analysis of Kim et al. [6] who evaluated performance in the gas sector through the use of the Edgeworth index, although their work concentrated on the segment of transport. The research presented here focuses on the retail sector since this is the area which has been most affected by the process of liberalization in Italy.

In order to explain the Edgeworth index, firstly the performance of two utilities (l and k) is considered (bilateral comparison). The utility l (l=1,2,...,L) consumes a varying amount of n inputs (x_n^l, n=1,2,...,N) to produce m different outputs (y_m^l, m=1,2,...,M). Analogously, the utility k (k=1,2,...,M) also consumes a varying amount of n inputs (x_n^k, n=1,2,...,N) to produce m different outputs (y_m^k, m=1,2,...,M). Input and output prices are denoted by w_n^l and p_m^l, respectively for utility l, and by w_n^k and p_m^k respectively for utility k. The Edgeworth index is used to calculate the difference between the output price and the input price of the two utilities in absolute values. This index employs the output quantities of the two companies as weightings for input and output price aggregation, as follows:

$$P_k^l = \sum_{m=1}^{M} \frac{\left(y_m^k + y_m^l\right)}{2} p_m^l - \sum_{m=1}^{M} \frac{\left(y_m^k + y_m^l\right)}{2} p_m^k \qquad (1)$$

$$W_k^l = \sum_{n=1}^{N} \frac{\left(x_n^k + x_n^l\right)}{2} w_n^l - \sum_{n=1}^{N} \frac{\left(x_n^k + x_n^l\right)}{2} w_n^k \qquad (2)$$

Price differential PW_k^l, which represents the effect of price on profit, can be defined as the difference between the output and input price, as follows:

$$PW_k^l = P_k^l - W_k^l \qquad (3)$$

This value shall be referred to from here on as 'price effect'.

The effect of production efficiency is calculated by subtracting the input from the output quantity difference as shown below.

$$\pi_k^l = \left(\sum_{m=1}^{M} \frac{\left(p_m^k + p_m^l\right)}{2} y_m^l - \sum_{n=1}^{N} \frac{\left(w_n^k + w_n^l\right)}{2} x_n^l \right) - \left(\sum_{m=1}^{M} \frac{\left(p_m^k + p_m^l\right)}{2} y_m^k - \sum_{n=1}^{N} \frac{\left(w_n^k + w_n^l\right)}{2} x_n^k \right) \qquad (4)$$

This value shall be referred to from here on as 'productivity effect'.

The variation in profit depends on the price difference and on the productivity (eqns 3 and 4).

The bilateral system does not satisfy the transitivity condition which is particularly important when comparing more than two utilities or performing a panel consisting of cross-sectional data.

Caves et al. [14] suggest that the transitivity condition could be met by introducing a hypothetical utility *h*. This utility *h* would have all its data equal to the sample means, and would then be used as a benchmark against which to measure all the other companies.

3 Data set

The data set, which was supplied by the Unione Italiana delle Camere di Commercio (Italian Union of the Chambers of Commerce), is comprised of 100 companies operating in Italy and includes their balance sheets.

The cross-sectional analysis which was performed refers to 2005. Since most of the retail companies are currently in the start-up phase, it was not possible to carry out an analysis which took series of historical data from the balance sheets into consideration. In fact most of the companies studied were formed between 2003 and 2005, which is too short a time period to obtain significant data.

On 31 December 2005, there were a total of about 400 companies authorised by the 'Ministero delle attività produttive' (Ministry of Productive Activities) to practise retail activities in the end market, but according to research undertaken by the Italian Regulatory Authority for Electricity and Gas (AEEG) only 257 of these companies appear to be active. Therefore our sample, which includes approximately 39% of the companies that are actually in business, is in fact a significant amount of the total number of retail companies.

The companies analysed have been divided according to their revenue into three groups: small, medium and large companies. 'Small' companies are considered to be those with an income of less than €10 million; 'medium' are those with an income of between €10 million and €50 million; and 'large' those which have an income which is greater than €50 million. Thus our sample is comprised of 38 small companies, 38 medium companies and 24 large companies.

It is assumed that the output of the natural gas retail sector is the quantity of gas sold, and that the main inputs are raw material (methane gas); labour; and indirect expenses/overheads, such as administrative costs which would include telephone calls, postage, publicity and sponsorship.

In order to apply Edgeworth's multilateral model, a hypothetical 'ideal company' *h* must be identified for every group, which shall then be used as a benchmark against which to measure the other companies [14]. Table 1 below shows these reference values for the benchmark company *h* in each group.

Table 1: Input and output values for company *h*.

Company *h*	**Output**		**Input**		
	Gas sold (Million m^3)	Profit (Thousands of euros)	Total Costs (Thousands of euros)	Gas purchased (Million m^3)	Number of Clients
Small company	12.45	4,118	4,236	12.54	1,798
Medium company	72.5	24,166	24,409	73.7	29,650
Large company	1,038	3,617,000	839,000	1,636	676,248

4 Analysis of the results

Through the use of the Edgeworth index it was possible to identify the way in which each company differed from the ideal company (*h*) in terms of: input costs, output costs, price differential, productivity, and profit. After having compared all the results for the companies, a graph of dispersion was plotted for each group in which the productivity effect was compared with the price effect. The gap between each company and the ideal, and the area where companies have a positive profit change may be seen from the graphs below.

In figure 1, the x-axis represents the productivity effect and the y-axis denotes the price effect. As stated above, the profit change is given by the sum of the productivity effect and the price effect, and thus the area above the bisector indicates a positive profit change where the greatest profits are to be found in the first quadrant. Companies below the bisector have a negative profit change and in the third quadrant both their productivity differences and price differences are negative.

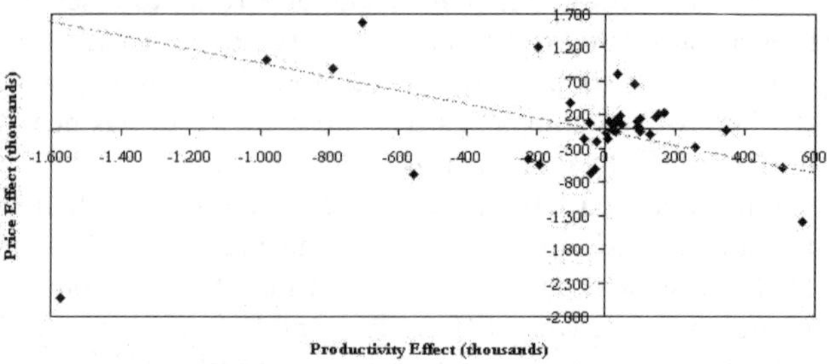

Figure 1: Performance trend of small companies.

As may be seen from figure 1, the small companies are very concentrated around the group average. There are small differences in profit and in price differential with respect to company *h*. This result may be due to the homogeneity of the organizational structure which typifies the small companies. Traditionally the small gas retail companies in Italy have always operated at a mainly local or provincial level, representing real local monopolies. Thus the only distinguishing factor between them is that of their geographic location.

In the analysis of the medium and large companies, there is a further subdivision to be considered regarding the strategy implemented following the liberalization of the sector. Many of the companies in these two groups have chosen a diversification strategy, operating not only in the field of gas retail, but also in those of the retail of other services, such as electricity and water.

The results suggest that as the dimensions of the businesses increase, the more companies tend to use diversification strategies. Indeed 67% of the large companies also compete in other sectors, a proportion which falls to 34% when referring to the medium companies. And only 5% of the small companies have diversified their businesses, which is the reason that they have not been considered in this analysis.

Figure 2 examines the performance trend of the medium companies. Firstly, it should be noted that with respect to the bisector there is less concentration around the average compared with the degree of concentration of the small companies. Moreover it can be seen that most of the medium companies are found above the bisector and thus have a positive profit change. In reality, such a result is due to the high level of productivity efficiency attained by medium companies, rather than being due to the mark-up achieved by these companies.

Most of the medium multi-business companies are positioned above the bisector, although many of these are found in the fourth quadrant where the mark-up obtained is below average.

In figure 3, the productivity effect and the price effect for large companies are compared. Firstly, it may be observed from the graph that there is a greater concentration of large companies in comparison to that of the medium companies. In relation to profitability, it should also be noted that approximately half of the companies manage to obtain a positive profit change and that this result is due to an increased mark-up rather than a greater level of productivity.

The large multi-business companies are almost equally divided between the area above and below the bisector.

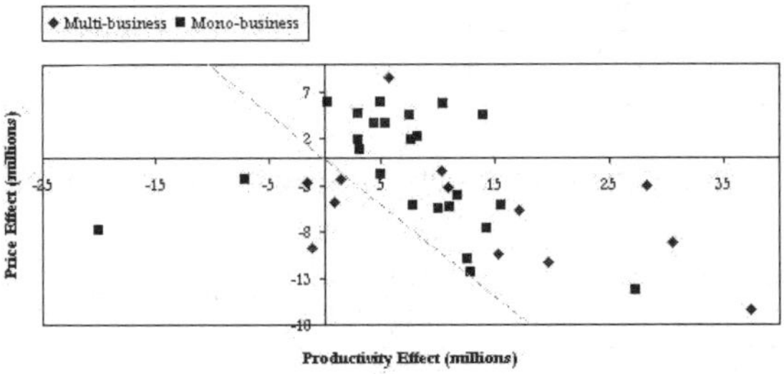

Figure 2: Performance trend of medium companies.

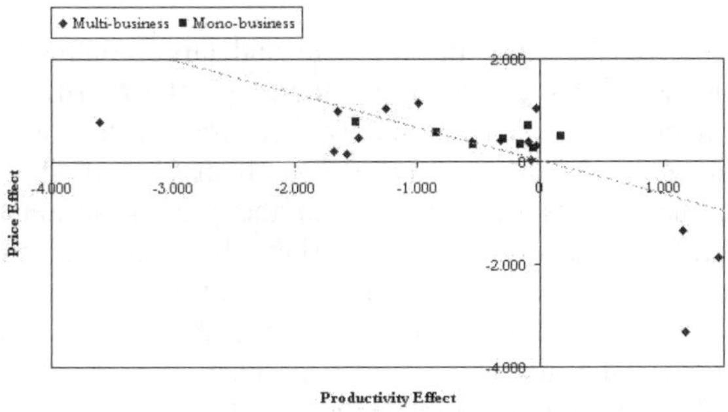

Figure 3: Performance trend of large companies.

5 Comparative analysis

It is interesting to analyse business performance in relation to company size and strategy adopted (diversification/specialisation).

The main performance results for each group, summarised in table 2, indicate that the most profit change is obtained by the medium sized companies, whilst the large companies show the least profit change. The high performance level of the medium companies is due to their greater productivity rather than to a larger mark-up. The opposite is true of the large companies which, thanks to their market power, manage to establish a higher mark-up. In fact 87% of the large companies boast a higher than average price differential owing to their greater bargaining power which allows them to purchase gas at lower prices from the suppliers. Despite such bargaining power, only 46% of the large companies have a positive profit change on account of the low level of productivity of the companies analysed. In fact only 17% of the large companies manage to attain a positive productivity change, a result which highlights an inefficient and wasteful allocation of resources.

In table 3, the multi-business companies and mono-product companies are compared in an attempt to determine whether the strategies adopted by the companies have a positive effect on profitability and productivity.

Table 2: Comparison of results for small, medium and large companies.

Group	Δ profits>0	Price differential>0	Δ productivity>0
SMALL	63%	42%	63%
MEDIUM	84%	53%	89%
LARGE	46%	87%	17%

Table 3: Comparison of results for specialised/diversified companies.

Group	**Strategy adopted**	**Δ profits>0**	**Price differential>0**	**Δ productivity>0**
MEDIUM	Specialised	92%	52%	92%
	Diversified	69%	8%	85%
LARGE	Specialised	63%	100%	13%
	Diversified	37%	69%	19%

In general the results from the medium and large companies indicate that those companies which have specialised attain a greater profit change than those which have diversified. This is mainly because on the whole specialised companies manage to obtain a greater mark-up than diversified companies. Such a disparity in mark-up is possibly due to the pricing strategies adopted by companies which decide to enter new markets. By reducing prices in the core business segment, the diversified companies offer clients an incentive to subscribe to their other utilities. The main aim of promotional initiatives of this nature is to subsidise the supply in the new markets.

With regard to productivity, the results show no significant difference between the specialised companies and the diversified companies in either group. Regardless of whether the companies are mono-business or multi-business, there is a low proportion of large companies which have attained a high level of productivity. Low levels of productivity in multi-businesses would appear to be counterintuitive since theoretically diversified companies ought to obtain a higher level of productivity thanks to the economies of scope. However in practice, given that the process of diversification has begun relatively recently, it is possible that more rewarding effects of this strategy have yet to mature.

6 Conclusions

This paper analyses the natural gas retail market following the relatively recent liberalisation of the sector.

In comparison to other sectors, the retail sector has been the most affected by liberalisation, and consequently many retail companies have used mergers in order to consolidate their market position. Further strategic options which were implemented after the market liberalisation were the specialisation of the core business or else diversification through entry into other markets such as those of electricity or water.

Following the liberalisation of the sector and the resulting tariff regulations it was interesting to analyse the market performance of the companies involved in gas retail, whilst considering the strategies used by these operators (specialisation/diversification). This analysis was carried out by applying the Edgeworth index, which enabled the integration of all three of the most important measures of strategic performance: profit, productivity and price differential.

Performance was studied by dividing the companies in the sample into three groups (small, medium and large). The data means of the companies in each group were used to define a hypothetical 'ideal company'. The companies in each group were then measured against our ideal company benchmark.

From our research it transpired that the medium sized companies have the best performance with respect to our ideal company in terms of profitability and productivity. In contrast the large companies mainly obtain positive profits due to an increased mark-up. On account of their greater bargaining power over suppliers, the large companies are able to purchase gas at lower prices.

In relation to the adopted strategies it is interesting to note that the specialised companies obtain a greater profit change with respect to diversified companies. Such a result may be due to the fact that the multi-utility companies are still in the start-up phase. Having to sustain greater reorganization costs, the diversified companies have not yet been able to take advantage of the economies of scope.

References

[1] European parliament and council *Directive 98/30/EC* of 22 June 1998 concerning common rules for the internal market in natural gas.

[2] European parliament and council *Directive 2003/55/EC* of 26 June 2003 concerning common rules for the internal market in natural gas and repealing Directive 98/30/EC.

[3] Capece, G., Di Pillo, F., Gastaldi, M., Levialdi, N., The European gas market: the effects of liberalization on retail prices, *Energy and Sustainability*, WIT Press, Volume 105, pp. 417–426, 2007.

[4] Ministry of Productive Activities *Legislative Decree n. 164* of 23 May 2000, concerning the implementation of the *Directive 98/30/EC* and the common rules for the internal market in natural gas.

[5] Capece, G., Di Pillo, F., and Di Stefano S., La performance del mercato della vendita del gas naturale in seguito alla liberalizzazione del settore, *Technical Report*, Department of Industrial Engineering, University of Rome "Tor Vergata", 21st January 2008, pp. 1–35.

[6] Kim, T-Y., Lee, J-D., Park, S-B., Profit, productivity, and price differential: an international comparison of the natural gas transportation industry, *Energy Policy*, Volume 27, pp. 679–689, 1999.

[7] Aivazian, V.J., Callen, J.L., Chan, M.W.L., Mountain, D.C., Economics of scale versus technological change in the natural gas transmission industry, *The Review of Economics and Statistics,* Volume 69, Issue 3, pp. 556–561, 1987.

[8] Sickles, R.C., Streitwieser, M.L., An analysis of technology, productivity, and regulatory distortion in the interstate natural gas transmission industry:

1977-1985, *Journal of Applied Econometrics,* Volume 13, Issue 4, pp. 377–395, 1998.

[9] Rushdi, A., Productivity changes in the gas and fuel corporation of Victoria, *Energy Economics*, Volume 16, Issue 1, pp. 36–45, 1994.

[10] Price, C.W., Weyman-Jones, T., Malmquist indices of productivity change in the U.K. gas industry before and after privatization, *Applied Economics*, Volume 28, Issue 1, pp. 29–39, 1996.

[11] Miller, D., Profitability + Productivity = Price Recovery, *Harvard Business Review*, Volume 3, pp. 145–153, 1984.

[12] Banker, R.D., Chang, H.H., Majumdar, S.K., A framework for analyzing changes in strategic performance, *Strategic Management Journal*, Volume 17, Issue 9, pp. 693–712, 1996.

[13] Erbetta, F., Fraquelli, G., Produttività e redditività nella distribuzione locale di gas naturale in Italia: proprietà, diversificazione e scala operativa, *L'Industria,* Volume 4, pp. 745–768, 2003.

[14] Caves, D.W., Christensen, L.R., Diewert, W.E., Multilateral comparison of output, input and productivity using superlative index number, *Economic Journal,* Volume 92, pp. 73–86, 1982.

A case study: the approach to the integrated and cooperative management of the water resources of the Maputo River Basin by Moçambique, Swaziland and South Africa

A. Tanner[1], D. Mndzebele[2] & J. Ilomäki[3]
[1]*Ninham Shand (Pty) Ltd, South Africa*
[2]*Ministry of Natural Resources and Energy, Swaziland*
[3]*Plancenter, Finland*

Abstract

The Maputo River basin, has its headwaters in South Africa, covers a very significant part of Swaziland and flows into the Indian Ocean, south of Maputo, in Moçambique. The Governments of Moçambique, South Africa and Swaziland have been collaborating in the exchange of information, agreements on sharing of water, and joint studies since the early 1980s. The Tripartite Permanent Technical Committee (TPTC), responsible for providing advice to the three watercourse States, was established in 1983. In 2002 the Interim IncoMaputo Agreement was signed and identified that a "Comprehensive Agreement" is required in order for the states to collaborate in the equitable utilisation, management, development and protection of the shared watercourses.

This paper describes the approach followed and the "Joint Maputo River Basin Studies" (JMRBS), which is being undertaken to establish the Comprehensive Agreement. The "Scoping Study", completed in 2005, identified, collected and reviewed the available data for the catchment, and recommended the scope and level of detail for the Integrated Comprehensive Study of the Water Resources of the Maputo River Basin.

The Assessment of Water Resources and Related Social and Economic Aspects of the JMRBS are currently underway. The objectives are to develop and model the socio-economics and water resources of the catchment to enable **Integrated Water Resources Management Scenarios (IWRMS)** to be proposed. It will support the three countries in their efforts to reach agreement on: sustainable development, alignment of objectives, coordination of management, sharing of benefits of the resources, while respecting the sovereignty of each country regarding its policies and subjects.

Keywords: shared watercourses, integrated water resources management, water resources, water quality, environmental flow requirements, economics, hydrology management and development options.

WIT Transactions on Ecology and the Environment, Vol 108, © 2008 WIT Press
www.witpress.com, ISSN 1743-3541 (on-line)
doi:10.2495/EEIA080061

1 The Maputo River Basin

1.1 Introduction

The Tripartite Permanent Technical Committee (TPTC) was formally established on 17 February 1983 to inter alia make recommendations to the three countries, namely the Kingdom of Swaziland, the Republic of Moçambique, and the Republic of South Africa (RSA), on any measures to be undertaken to alleviate short-term problems regarding water shortages on all rivers of common interest during drought periods, on the division of water allocations, joint schemes, arrangements for the investigation of an access to common watersheds and joint water schemes and the required mechanism to co-ordinate and integrate the findings and plans of each country, including the appointment of Consultants as required.

It was clear that all three countries would benefit if, among other river basins, the water resources of the Maputo Basin were managed and developed in an integrated manner. Because the river basin extends across the political and administrative boundaries of Moçambique, Swaziland and South Africa, integrated management requires joint management.

1.2 The interim agreement

Various studies of the tributary catchments of the Maputo River have been conducted by the three countries and the information has formed the basis for an Interim Agreement (TPTC [1]) between the Republic of Moçambique, the Republic of South Africa and the Kingdom of Swaziland for co-operation on the protection and sustainable utilisation of the water resources of the Incomati and Maputo Watercourses. This Interim Agreement was formalised at the World Summit on Sustainable Development, which was held in August and September 2002 in Johannesburg.

However, a Comprehensive Agreement for the Use and Management of the Shared Waters of the Maputo Basin, the "Comprehensive Agreement" is required for the joint development and management of water resources by the three countries sharing the Maputo River Basin, in order to:

a) Enhance effective utilisation of the water resources of Maputo River;
b) Enhance sustainable economic development; and
c) Enhance the quality of the natural and social environments of the Maputo River Basin.

1.3 Catchments and climate of the basin

1.3.1 Catchments

The Maputo River Basin is located on the eastern side of the great continental divide. The basin is bordered by the Umbeluzi and Incomati Basins to the north, and the Umhlatuze coastal catchment to the south. The land area of the basin is about 30 000 km^2. The headwaters of the basin originate in South Africa, and the main tributaries flow through the southern half of Swaziland (Usuthu) and further south through South Africa (Phongolo), before joining on the South Africa / Moçambique border and flowing to the estuary in Maputo Bay (Figure 1).

The catchments of the Usuthu, Phongolo, and Ngwavuma Rivers, and the Maputo River downstream of the confluence of the Usuthu and Phongolo Rivers form the main sub-basins of the Maputo River Basin. The Usutho sub-basin is comprised of the Lusushwana, Mpuluzi, Usuthu (main stem), Ngwempisi and Mkhondvo catchments. The description of the Maputo River Basin in the Interim IncoMaputo Agreement is based on this basin subdivision.

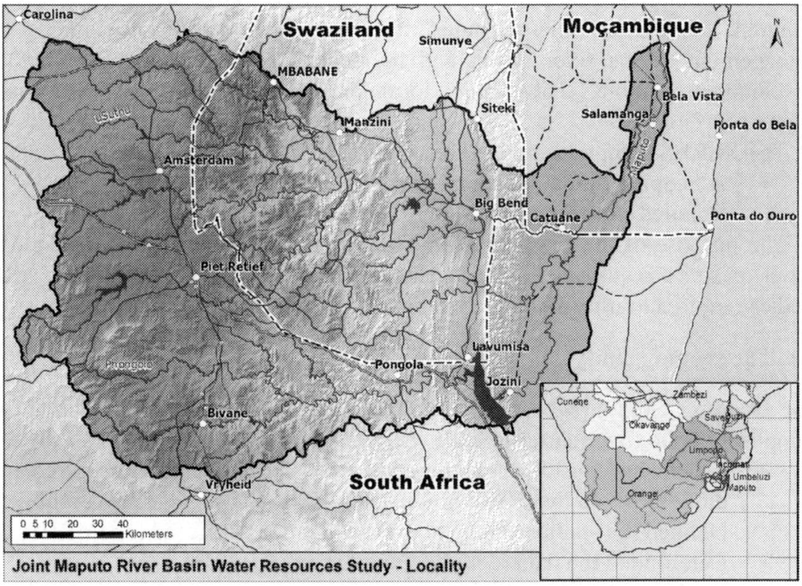

Figure 1: Locality map.

1.3.2 Climate

The climate of the Maputo River Basin is for the most part sub-tropical, but varies considerably from west to east. In the west, the climate of the interior plateau is temperate and sub-humid, the Lowveld is semi-arid and warm, and the Coastal Plain is warm to hot and humid. Along the south eastern coast of Africa, warm, moisture laden air of the subtropical eastern maritime system moves in from the sea. When the moist air is forced to rise above the land mass, cooling, condensation and rainfall occurs. The moist, warm air mass is unstable, and is the source of cyclonic activity. Very few cyclones make landfall, but when this happens, devastating floods result, such as those caused by cyclone Domoina in January 1984, Imboa in February 1984, and Eline in February 2000. The tropical cyclone season lasts from November to April, with activity peaking in January and February.

Mean annual rainfall over the Maputo River Basin vary from about 500 mm/year in the Makhathini Flats on the Coastal Plain, to about 1400 mm/year in the vicinity of Mbabane on the Escarpment. Mean annual evaporation is lowest on the cooler interior plateau (1300 mm/year) and highest in the Coastal Plain, roughly inversely proportional to rainfall (MNRE [2]). Larger deficits in crop water requirements can therefore be expected in the eastern parts of the basin.

2 The joint Maputo River Basin studies

2.1 Introduction

The three countries agreed that a Joint Maputo River Basin Study (JMRBS) was required to provide the technical information about the resources and management of the river basin, and to identify alternative management and development options that would form the basis for the "Comprehensive Agreement".

The JMRBS comprises two phases:

- A Scoping Study, which has been completed; and
- Detailed studies.

The purpose of the Scoping Study was to establish the level and additional depth of study required and to develop a Study Plan for the detailed JMRBS studies required to inform the preparation of a Comprehensive Agreement.

2.2 The scoping study

The Scoping Study (MNRE [3]) commenced in November 2003 and was completed in the first quarter of 2005. It:

- Established the issues to be addressed and deliverables required from the JMRBS so that it can provide a reliable source of information;
- Defined the studies and level of investigations required for JMRBS;
- Determined the data required to perform the JMRBS;
- Established the availability, quality, age, resolution, compatibility, homogeneity and adequacy of the data, and documented the status of the data;
- Established the extent of the additional data which must be obtained, and the areas where the quality of the existing data must be improved;
- Completed an initial stakeholder analysis, which included goals and objectives for Public Participation (PP), participation mandates, participation objectives, stakeholder profiles, and stakeholder organization and representation;
- Held a series of stakeholder workshops, which explored, discussed and agreed Basin Study objectives, data sources and gaps; and
- Developed relationships between relevant stakeholder groups.

2.3 The comprehensive studies

The objectives of the comprehensive JMRBS are to provide a detailed water resource assessment and recommended management and development options, as well as to propose suitable institutional and financial arrangements for the Maputo River Basin. These will be presented as an Integrated Water Resources Management Strategy (IWRMS).

The objective of the IWRMS is to provide a framework for the planning, development and management of a sustainable water resources system that can be used to fully contribute to the improvement of social and economic objectives of the three Basin States, while maintaining the water resources on a sustainable, high quality basis and avoiding degradation of the natural environment of the Maputo River Basin. The results of the JMRBS and the IWRMS will form the basis for the "Comprehensive Agreement".

During the Scoping Phase, the following study modules were identified for the second phase of the JMRBS.

2.3.1 Study management

The objective of the Study Management component of this assignment is to provide effective technical assistance to the TPTC Task Team with the procurement, management and co-ordination of the human, material and financial resources to achieve the objectives of the technical components of the JMRBS and, if required, to formulate technical sections of the Comprehensive Agreement for the Maputo River Basin.

2.3.2 Public participation

The objective of the Public Participation task is to ensure extensive consultation and participation throughout in the JMRBS, and to lay the foundation for ongoing and effective participation under the Comprehensive Agreement. Specific objectives are:

- Creating awareness amongst stakeholders that will enable any of the Study Teams to interact effectively with a range of stakeholders.
- Gaining an adequate and consistent understanding of stakeholders concerns and expectations.
- Providing a legitimate and effective consultative base for cooperative basin management, this will increase the probability that the Comprehensive Agreement can be implemented with stakeholder buy-in.
- Development of an environment of understanding of and support for the proposed "Comprehensive Agreement".
- Consolidating existing Basin State capacity and processes for PP.
- Security a resource and tool for participatory IWRM under the Comprehensive Agreement.

2.3.3 Assessment of legal, institutional and financial arrangements

The legal, institutional and financing consultancy will provide the legal, institutional and financial background to enable the Comprehensive Agreement to be implemented in practice. The aim of the Legal, Institutional and Financing consultancy is to recommend an institutional framework for cooperative management of the water resources. The proposals need to be practical and capable of implementation within the specific institutional and legal environments pertaining in each country.

2.3.4 Environmental water requirements

The Environmental Water Requirements (EWR) of the Maputo River and estuary are likely to be one of the most significant factors in determining the available yield and the potential for further water resource development. Furthermore, the EWR is essential to inform the management of the system in a sustainable manner and to predict the likely consequences of any change in flow regime. The objectives are to determine the EWRs for the ecological and social environments. This will include the temporal and spatial distribution of flows, as well as the water quality objectives for these environments.

The current differences between the countries with respect to the implementation and determination of EWRs and the methodologies used for assessing EWRs create challenges in terms of this assignment.

Acknowledging the holistic nature of the approach, it is important to recognize the potential for cumulative impacts on the Maputo Bay.

2.3.5 Assessment of the water resources and related social and economic aspects of the maputo river basin

This consultancy, now known as the "Joint Maputo River Basin Water Resources Study", (JMRBWRS) is currently underway and is discussed in Section 3.

3 The joint Maputo River Basin water resources study

In February 2006 the Government of Swaziland, acting on behalf of the European Commission and the three Countries, appointed consultants to undertake the JMRBWRS. The Study is being undertaken by Consortium consisting of Plancenter (Lead, Finland) and three South African Consulting Companies: Ninham Shand Consulting Services, Diversity and Transformation Solutions and Water for Africa. The end date of the Study is scheduled for July 2007. The objectives are to provide a detailed water resource assessment and recommended management and development options for the Maputo River Basin.

It is assessing the socio-economic and water resources information for the catchment to enable Integrated Water Resources Management Scenarios (IWRMS) for the optimum and sustainable utilisation of resources of the catchment to be prepared. Different development scenarios are being analysed with the associated environmental, socio and economic impacts and mitigation measures. It will enable decision-makers to reach agreement on the sharing of benefits of the resources of the catchment and move towards an integrated approach to the management of the system.

The initial findings are described in the following sections.

4 Water use and water resources

4.1 Water use

The major water users in the Maputo Basin are irrigation, afforestation and transfers from the Upper Usuthu catchment to the Vaal and Olifants catchments.

Irrigated agriculture and commercial forestry the largest water users in the basin, have expanded by about 1.4% and 0.6% per annum on average over the past 25 years, and now cover about 55 000ha and 367 000ha respectively. Irrigated agriculture in the Moçambique portion of the basin has declined from about 3800 ha in 1981 to about 200 ha in 2005.

There are five areas in the basin where formal irrigation schemes have been established for smallholder farming, totalling about 16 000 ha in Swaziland (after completion of the LUSIP scheme), 4000 ha in South Africa, and about 180 ha near Salamanga in Moçambique. Plans have been put forward to expand most of the smallholder schemes.

There are large areas of dry land subsistence farming in the lower Usuthu and Ngwavuma catchments, and smaller areas in the Ngwempisi, Lusushwana and Mkhondvo catchments. Individual land holdings are generally less than 2 ha in size. Subsistence farming areas in the lower Phongolo and Maputo catchments are mostly located in the floodplain, and are usually irrigated with furrows.

Available data indicate that at least 24 000 ha (equivalent area at 100% density) of the Maputo River Basin land area has been invaded by alien plant species. Compared to the rest of the basin, the Maputo catchment in Moçambique is relatively free of invasive alien plants (MNRE [4]).

4.2 Groundwater aquifers

The eastern half of the Coastal Plain is situated on the largest primary aquifer in Southern Africa. Along the Maputo River, the unconsolidated alluvial terraces have some of the highest yields in the basin (>5 l/s) (MNRE [5]). It is very likely that there is significant interaction between the porous alluvial aquifers and the river channel, implying that the groundwater and surface water in these areas should be managed as one resource. Aquifer yields decline in the semi-consolidated Cretaceous rocks in the western half of the Coastal Plain. There is a small, but high yielding (2 – 5 l/s) karst aquifer located upstream of the confluence of the Phongolo and Ngwavuma Rivers, between the two rivers. Aquifers in the Lebombo Range and Lowveld have some of the lowest yields in the basin (0.1 – 0.5 l/s). The Lebombo Rhyolites are generally fresh, hard, and non-water bearing. Local fracturing of the massive basalt rocks do however provide water-bearing zones with recorded blown yields of up to 7 l/s. Metamorphic units (mainly gneisses) are widely distributed in the Escarpment Region. The gneisses are not significantly weathered, and are therefore not easily infiltrated. The granitic areas in the Middleveld, Escarpment and Interior Plateau regions are favoured targets for groundwater exploration. Aquifers are moderately productive, with yields of about 0.5 to 2 l/s. There are also contact springs in the area.

4.3 Current surface water resources

4.3.1 Quantity

The Maputo River Basin comprise of eight catchments and covers about 29 970 km^2. The natural mean annual runoff (MAR) amounts to about 3 770 Mm^3/a, of which about 58% is from South Africa, 40% from Swaziland and 2% from Moçambique (MNRE [6]).

It is estimated that the development in the Maputo Basin has reduced the natural MAR by between 25% and 30%, while the runoff in the Usuthu and Phongolo Rivers upstream of Moçambique have been reduced by about 25 and 35%, respectively.

4.3.2 Quality

In terms of its fitness for use, potential problems relate to the build-up of salts, potential nutrient enrichment problems, and potential problems related to acid mine drainage or acid rain deposition (MNRE [7]).

4.4 Water resources infrastructure

Most of the water infrastructure in the Maputo River Basin has been developed during the past 50 years. There are two large schemes that transfer water from the upper Mkhondvo and Ngwempisi catchments to the Vaal system for power generation; regional irrigation schemes in the lower Usuthu, Ngwavuma and Phongolo River catchments; and smaller local schemes for water supply to the towns and rural settlements in the basin (MNRE [2]).

4.5 Water balance assessments

An initial water balance has been completed to identify sub-systems with development potential, as well as areas where deficits are imminent. The water balance calculation was performed by Water Management sub catchment. Analysis was for a present day (2005) scenario.

The water balance exercise found that there is surplus water available in the Upper Phongolo catchment, as well as from the Bivane and Pongolapoort Dams. There is a shortfall at the dams in the Upper and middle Usuthu catchments, which is mainly as a result of the transfers, combined with the potential environmental requirements for the areas downstream of these dams (MNRE [8]).

5 Future scenarios

5.1 Introduction

An assessment of local and regional spatial development plans, the probable EWRs, the initial water balance, projected future water use as well as management and infrastructure development options, shows that a likely short to medium term scenario could look as follows (MNRE [2]):

Swaziland:

- Significant expansion (+-14 000 ha) of smallholder irrigation in the Lusushwana, Mkhondvo, lower Usuthu and Ngwavuma catchments;
- Limited expansion of commercial plantations in the escarpment region;
- Development of small dams and groundwater resources to improve assurance of existing supplies;
- Improved efficiency of water use in the irrigation sector;
- Commencement of a programme to eradicate invasive alien plants;
- Development of one or two major dams to support the increased areas of irrigation and further improve assurance of supply, while meting the EWR.

Moçambique:

- Expansion of smallholder and commercial irrigation, mainly through the rehabilitation of disused irrigation schemes in the Salamanga and Catuane area;
- Some expansion of commercial plantations (outside the boundaries of the Maputo water course) due to renewed foreign investment in this sector.
- Possible water transfers for urban use (Ponta Dobela and/or Maputo)
- Development of one or possibly two off channel storage dams to improve the assurance of supply to existing and future users while meeting the EWR.

South Africa:

- Very limited expansion of irrigated areas, almost exclusively to benefit rural poor, and making use of existing unused water allocations;
- No significant expansion of commercial plantations;
- Some expansion of dry land (sugar cane) agriculture;
- Limited development of small dams and groundwater resources to improve assurance of supply;
- Increased municipal (domestic) water use as water and sanitation service backlogs are gradually reduced. Improved efficiency of water use in the irrigation sector;
- Improved efficiency of water use in the irrigation sector;
- Continued progress with eradication of invasive alien plants; Maintenance of existing inter-basin transfers
- Development of one or two major dams to maintain the assurance of supply for existing transfers and water users while meeting the EWR.

6 Conclusions

The Study is now at the point where all the key information and basic analyses have been completed. The development of the IWRMS for consideration by the three countries is in progress. The challenges are; improve the confidence in the water resource assessment through further monitoring and calibrations, to improve the confidence in the EWRs through more detailed studies, improve the efficiency of water use and to improve the socio economic benefits of water used in the catchment.

Acknowledgements

The authors gratefully acknowledge the study funding provided by the EU, the support of the officials from all their countries and the contributions of all members of the study team to the ongoing studies.

References

[1] TPTC – Tripartite Technical Committee, for Moçambique, South Africa and Swaziland, Interim Inco Maputo Agreement, 2002.
[2] MNRE – Ministry of Natural Resources and Energy, Swaziland, representing DNA of Moçambique and DWAF of South Africa, Report on Basin Characteristics, Land Use and Water Resources Infrastructure, 2007.
[3] MNRE – Ministry of Natural Resources and Energy, Swaziland, representing DNA of Moçambique and DWAF of South Africa, Report on Integrated Scoping Phase of the Water Resources of the Maputo River Basin, 2007.
[4] MNRE – Ministry of Natural Resources and Energy, Swaziland, representing DNA of Moçambique and DWAF of South Africa, Report on Baseline Surface Water Resource Availability, 2007.
[5] MNRE – Ministry of Natural Resources and Energy, Swaziland, representing DNA of Moçambique and DWAF of South Africa, Report on Ground Water Resources, 2007.
[6] MNRE – Ministry of Natural Resources and Energy, Swaziland, representing DNA of Moçambique and DWAF of South Africa, Report on Present Water Resource Availability, 2007.
[7] MNRE – Ministry of Natural Resources and Energy, Swaziland, representing DNA of Moçambique and DWAF of South Africa, Report on Surface Water Quality Assessment, 2007.
[8] MNRE – Ministry of Natural Resources and Energy, Swaziland, representing DNA of Moçambique and DWAF of South Africa, Report on Water Requirements and Assessment Water Resource Availability Water Balance, 2007.

Section 2
Environmental policies, planning and assessment

Assessing the external costs and the economic viability of the Greek steel industry

D. Damigos & D. Kaliampakos
School of Mining and Metallurgical Engineering,
National Technical University of Athens, Greece

Abstract

In recent years, legislative requirements and environmental policies at European, as well as national, level seek to internalize the environmental impacts that have been traditionally viewed as externalities, in order to come up with more informed and fair choices. The IPPC (Integrated Pollution Prevention and Control) Directive 96/61/EC lays down a framework requiring Member States to issue operating permits for certain installations based on best available techniques (BAT) in order to achieve a high level of protection of the environment. This framework gives clearly importance to economic aspects. More specific, the environmental effects of an installation or a sector are compared against the costs for taking preventive measures against pollution, in order to determine which, if any, meet the criteria of BAT. The scope of the paper is to explore the effects of BAT implementation in the Greek steel sector towards eliminating air emissions from steel production. The analysis is based on pollutant emissions gathered by reports prepared for the European Pollutant Emission Register (EPER) and on external costs, in terms of euros per tonne of pollutant emitted, generated by European Programmes. The externalities estimated are compared to important financial indicators of individual steel producers and of the sector, as well, in order to provide necessary input for assessing the economic viability of the industry under investigation.
Keywords: externalities, steel industry, air emissions, best available techniques.

1 Introduction

It is commonly accepted that iron and steel, together with coal, have played an important role in the development of human civilisation. These were the

WIT Transactions on Ecology and the Environment, Vol 108,
www.witpress.com, ISSN 1743-3541 (on-line)
doi:10.2495/EEIA080071

principal materials upon which the industrial revolution was based, founding numerous uses in agriculture, construction, manufacturing of machinery, medicine, etc.

World crude steel production has grown exponentially in the second half of the twentieth century, rising from 189 million metric tons in 1950, to 848 million metric tons in 2000. Over the last decade, world crude steel production has grown by 56.5%, from 799 to million metric tons in 1997 to 1,250 in 2006. In the same period, EU-25 crude steel production has increased only by 7.5%, as shown in Table 1, and the European share of world crude steel production has steadily declined from 23.1% in 1996 to 15.9% in 2006 [1].

Table 1: Crude steel production 1997–2006 ('000 metric tons).

	1997	1998	1999	2000	2001	2002	2003	2004	2005	2006
EU-15	159,867	159,888	155,209	163,358	158,497	158,686	160,975	169,071	165,112	173,233
EU-25	184,568	182,424	175,943	186,694	180,546	180,896	184,505	194,189	187,213	198,462
Other Europe	26,367	25,449	22,390	23,707	24,604	26,615	29,288	32,139	33,186	36,378
C.I.S.	80,558	73,950	85,657	98,489	99,619	101,089	106,220	113,112	112,876	119,766
North America	129,489	129,945	130,044	135,353	119,858	122,949	126,161	134,021	127,631	131,655
South America	36,966	36,121	34,594	39,110	37,372	40,861	43,047	45,875	45,316	45,298
Africa	12,856	12,806	12,818	13,827	14,916	15,807	16,289	16,706	17,995	18,780
M.East	9,929	9,065	9,779	10,780	11,690	12,492	13,443	14,253	15,257	15,376
Asia	308,633	297,873	308,799	331,880	353,801	394,928	442,394	510,095	598,083	675,589
Oceania	9,589	9,697	8,946	7,832	7,859	8,292	8,397	8,300	8,646	8,691
World	798,954	777,330	788,970	847,671	850,266	903,929	969,743	1,068,691	1,146,203	1,249,997

Source: IISI.

Although investments and employment have diminished, steel industry remains a key sector for Europe's economy and competitiveness, since it accounts for about 1.8% of the value added and 1.5% of employment in EU manufacturing [2]. For this reason, the European Commission has been concerned about the crisis in the European steel industry and has aimed at stabilising the intra-Community steel market. Towards this direction the European Parliament, in its resolution of 12 February 2004, called for measures to be taken at Community level to defend Europe's iron and steel industry (i.e. regulation of unfair competition from outside the EU). Further, in its resolution of 24 February 2005, Parliament invited the Commission, after the expiry of the ECSC Treaty, to present a strategy for the future prospects of the steel sector in order to promote independent European capacity in this sector [2].

Nevertheless, it is known that EU policy recognizes that economic development must be sustainable with respect to the environment and from an environmental viewpoint, steel industry is an important emitter of air pollutants (i.e. dust, NO_x, SO_2, etc.) and CO_2 and is highly intensive in both materials and energy. During the last 20 years, the energy required to produce a tonne of steel has fallen by 40%, and throughout the nineties there has been a reduction of 20% in CO_2 emissions for the industry [2]. In addition, steel is the most recycled material in the world, since it is 100% recyclable with no downgrading in quality. It is estimated that about 47% of EU steel production is made from recycled scrap, while recycling of steel allows the saving world-wide of about 600 million tonnes of iron ore and 200 million tonnes of coke each year [3]. Yet, air pollution remains an important issue.

In some countries, the environmental pollution caused by steel production is a matter of trade union concern. For example, in Italy, the unions consider the introduction of pollution-control systems in pursuit of zero environmental impact to be of priority importance, asking the government to set up a permanent forum for discussion on industrial policies in the medium to long term [4]. In general, however, the steel industry is mainly affected by EU and national policies and measures. The EU environmental policy context, and consequently its adoption at national level, is driven by the increased prominence of sustainable development and it emphasizes in market-based instruments at ex ante (e.g. permission processes) as well as ex post procedures (e.g. liability).

Although there are several Directives affecting the steel industry, this paper emphasizes on the requirements set by the IPPC (Integrated Pollution Prevention and Control) Directive 96/61/EC [5]. More specific, the aim of this paper is to determine whether or not the implementation of a BAT package in pursuit of practically zero air emissions is considered to be viable for the Greek steel industry under the IPPC framework. The paper focuses on the external costs and the resilience of the individual companies and the sector as a whole, considering annual emissions of specific air pollutants and a number of financial ratios.

2 Evaluating the economic viability of a sector

The IPPC Directive lays down a framework requiring Member States to issue operating permits for certain installations, including steel units. The importance of this Directive consists in the fact that these permits must contain conditions based on best available techniques (BAT) to achieve a high level of protection of the environment. According to the Article 2.11 of the IPPC Directive, "best" means the most effective ones in achieving a high general level of protection of the environment as a whole and "available" means those techniques developed on a scale which allows implementation in the relevant industrial sector, under economically and technically viable conditions.

In order to assist in the determination of BAT under the IPPC Directive, the European IPPC Bureau (EIPPCB) organizes the exchange of information and produces BAT reference documents (BREFs), which Member States are required to take into account. Among others, the EIPPCB has prepared a horizontal BREF entitled "Economics and Cross-media Effects (ECME)" [6]. According to the methodology described in the abovementioned BREF, the selection of BAT under IPPC Directive takes into account the likely costs and benefits of pollution reduction measures as well as the results of an environmental cross-media assessment in order to avoid creating a new environmental problem when solving another. At the final step of the methodology, an evaluation procedure is set out in order to ensure that whichever technique is determined to be BAT does not undermine the economic viability of the industrial sector implementing that measure.

Determining whether implementing a BAT option in a sector is 'economically viable', depends on the capacity that the sector has to absorb the extra cost, or to transfer these costs on to the customer or suppliers [6]. The ability of the sector to absorb the costs depends on its resilience whereas the ability of the sector to pass the costs on depends on the structure of the industry and the market, as well.

The industry structure involves a number of sector's characteristics, such as the size and the number of installations, the production processes applied, the barriers that prevent players entering or leaving the market, etc. As far as the market structure is concerned, several factors influence the ability of the firms to pass on the costs of BAT implementation to the consumers, such as the extent of the market (i.e. local or global), the price elasticity of the commodities produced by the firms, the competition among existing firms and the threat of new entrants, etc.

The 'resilience' refers to the sector's ability to absorb the increased costs of environmental improvement in the short-, medium- and long-term. In order to ensure this viability, firms in the sector will need to be able to generate sufficient financial returns on an ongoing basis. There are several financial ratios that are used to describe the economic situation of a company. These financial ratios can be useful for evaluating company's resilience, but they can be difficult to apply to a sector. Hence, when carrying out the viability assessment, an 'average' (hypothetical) company can be used by averaging, for example, the annual accounts for the sample of the companies of the sector under investigation. However, the results can easily be distorted by the selection of companies in the sample. These distortions are more likely in cases where there are fewer companies in the sector or where there are some particularly badly or well performing companies.

3 Evaluating the economic viability of the Greek steel sector

3.1 Overview of the Greek steel industry

Steel sector plays an important role in Greek manufactory industry. The production of Greek steel industry increased by almost 140% over the last decade, that is 20 times more than the average increase rate in EU-25. Greece produced 2.4 million metric tons of crude steel in 2006 and was ranked 39th in the world.

Strictly speaking, the steel industry involves the production of crude steel, semi-products, hot-rolled finished products, continuously cast products, cold-rolled sheets and plates, and coated sheets. Nevertheless, in the analysis presented only those companies involved in crude steel production by scrap melting are examined.

One of the most important characteristics of the Greek steel industry, which is common in the majority of countries, is its high degree of concentration due to the increasing scale of production units as well as mergers and acquisitions between companies and groups. In Greece, only five companies account for the total of steel output and employment, namely:

- Halyvourgiki S.A.
- SOVEL S.A.
- SIDENOR S.A.
- Halyvourgia Thessalias S.A.
- Hellenic Halyvourgia S.A.

These five companies belong to three groups, i.e. Halyvourgiki, SIDENOR – SOVEL and Halyvourgia Thessalias - Hellenic Halyvourgia, which operate the following production units:

Halyvoyrgiki

Halyvourgiki's melt shop is located at Aspropyrgos, Attiki Region. Steel is produced in the form of square cross section prisms (billets). The basic production stages of the steel making process include melting of steel scrap, secondary metallurgy (fine adjustments to steel composition) and continuous casting of molten steel. The melt shop includes electric arc furnaces, with 100-ton capacity each, for the melting of steel scrap and a ladle furnace where secondary metallurgy takes place [7].

SIDENOR – SOVEL

The group operates two units at Thessaloniki (Northern Greece) and Almyros, Magnesia (Central Greece). As far as steel production is concerned, the Thessaloniki plant includes one 80 t electric arc furnace for melting of steel scrap and one 80 tn ladle furnace for secondary metallurgy. The Almyros industrial complex comprised of a steel plant, rolling-mill facilities, a construction mesh production unit, a pipe manufacturing unit and auxiliary units. Steel production from scrap takes place in one 130 tn electric arc furnace and one 130 tn ladle furnace [8].

Halyvourgia Thessalias - Hellenic Halyvourgia

The group owns two units at Aspropyrgos (Attiki) and Velestino, Magnesia (Central Greece). The Aspropyrgos industrial complex is comprised by a melt shop, a rolling-mill for long products, a wire mesh plant, as well as covered warehouses. The Velestino melt shop has an annual production capacity in semi-finished product (billet) that exceeds 700.000 tons. The industrial complexes include electric arc furnaces for melting of steel scrap and ladle furnaces for secondary metallurgy [9].

All the abovementioned facilities operate advanced control systems (i.e. fume and water treatments plants) and use BAT in order to minimize environmental impacts of steel production. In addition, the companies apply environmental management systems, according to the international standard ISO 14001:2004.

3.2 Assessing the externalities of the sector

In order to assess the externalities of the Greek steel sector in monetary terms for the purposes of this analysis, two types of information were considered, namely the annual emissions of the plants and the external cost of the pollutant expressed as €/tn emitted.

As far as the emissions are concerned, values declared by the firms for the European Pollutant Emission Register (EPER) report were used. It should be mentioned that according to the EPER Decision [10] the report covers 50 pollutants which must be included if the threshold values indicated in Annex A1 of the EPER Decision are exceeded. Given that in the vast majority of the cases the sector's emissions do not exceed the threshold values, EPER database does not include all the information needed; therefore the company statements for the 2004 emissions were used (Table 2).

Table 2: Steel sector's emissions in 2004 (kgr).

	NO_x	SO_2	PM10	NMVOC
SOVEL	56,000	365,000	20,700	0
SIDENOR	85,383	30,627	15,663	0
Halyvourgia Thessalias	37,800	10,210	6,760	36,110
Hellenic Halyvourgia	36,454	6,432	4,634	25,150
Halyvoyrgiki	25,000	44,928	0	0

Following the 'ECME' BREF guidelines, the external costs derived from the cost benefit analysis in the Clean Air for Europe (CAFE) programme, were applied [11]. Given that external costs have only been derived for a few air pollutants, namely $PM_{2.5}$, NH_3, SO_2, NOx and VOCs, values proposed for $PM_{2.5}$ and VOCs were used for estimating externalities of PM10 and NMVOCs, respectively.

It is underlined that the external costs in the CAFE programme relate only to human health and crop damages. Ecosystem externalities could not be monetised due to lack of data. In addition, many assumptions have been made, both when establishing the predicted environmental effects and when deriving values for the predicted impacts, which may push the results either way, up or down. Hence, it is recommended that ranges are used and sensitivities explored.

Table 3: Marginal damage (€ per tonne emission).

	Low value	Upper value
NO_x	840	1,900
SO_2	1,400	4,000
PM10	8,600	25,000
NMVOC	280	880

Table 4: Annual external costs (€) per pollutant of Greek steel industry.

	NO_x		SO_2		PM10		NMVOC	
	Lower	*Upper*	*Lower*	*Upper*	*Lower*	*Upper*	*Lower*	*Upper*
SOVEL	47,040	106,400	511,000	1,460,000	178,020	517,500	0	0
SIDENOR	71,722	162,228	42,878	122,508	134,702	391,575	0	0
Halyvourgia Thessalias	31,752	71,820	14,294	40,840	58,136	169,000	10,111	31,777
Hellenic Halyvourgia	30,621	69,263	9,005	25,728	39,852	115,850	7,042	22,132
Halyvoyrgiki	21,000	47,500	62,899	179,712	0	0	0	0
Sector	202,135	457,210	640,076	1,828,788	410,710	1,193,925	17,153	53,909

Bearing these remarks in mind, in order to estimate the externalities in the case of the Greek steel industry, the lower and the upper values suggested by CAFE for pollutants emitted in Greece were used (Table 3). The lower and upper estimates of the analysis are presented in Tables 4 and 5.

Table 5: Total annual external costs (€) of Greek steel industry.

	Total	
	Lower	*Upper*
SOVEL	736,060	2,083,900
SIDENOR	249,301	676,311
Halyvourgia Thessalias	114,293	313,437
Hellenic Halyvourgia	86,521	232,973
Halyvoyrgiki	83,899	227,212
Sector	1,270,074	3,533,832

3.3 Evaluating the economic viability of the sector

In order to examine the financial burden that will be placed on the steel industry and its firms from implementing technologies that will minimize air emissions, two measures were used, namely:

- Earnings before interest and taxes (EBIT)
- Earnings before taxes (EBT)

'EBIT' is regarded as the most appropriate indicator of operational performance because the comparisons are not influenced by the particular way that the company is financed. 'EBT' is an important profitability measure because deducts all expenses from revenue except from the payment of tax. Thus, it provides a good idea of fluctuations in companies' profits from year to year. In order to be consistent with the emission reference year, the profitability measures were collect from the 2004 financial statements (Table 6).

Table 6: .Profitability measures ('000 €) of Greek steel industry in 2004.

	EBIT	EBT
SOVEL	40,489	38,642
SIDENOR	28,543	20,790
Halyvourgia Thessalias	11,852	6,831
Hellenic Halyvourgia	3,729	2,134
Halyvoyrgiki	14,632	4,144

The external costs were expressed as a percentage of the above measures in order to form the 'externality' ratios. Although there is no pre-determined percentage of accepting the results, these values provide useful insights for assessing the economic viability of the sector; it is evident that firms with lower 'externality' ratios will find it easier to absorb the costs of implementing BAT. The results of the analysis are presented in Table 7.

Table 7: 'Externality' ratios (%) of Greek steel industry in 2004.

	SOVEL	SIDENOR	Halyvourgia Thessalias	Hellenic Halyvourgia	Haly-voyrgiki	Sector
NO_x						
%EBIT	0.12%	0.25%	0.27%	0.82%	0.14%	0.20%
upper	0.26%	0.57%	0.61%	1.86%	0.32%	0.46%
%EBT lower	0.12%	0.34%	0.46%	1.43%	0.51%	0.28%
upper	0.28%	0.78%	1.05%	3.25%	1.15%	0.63%
SO_2						
%EBIT	1.26%	0.15%	0.12%	0.24%	0.43%	0.64%
upper	3.61%	0.43%	0.34%	0.69%	1.23%	1.84%
%EBT lower	1.32%	0.21%	0.21%	0.42%	1.52%	0.88%
upper	3.78%	0.59%	0.60%	1.21%	4.34%	2.52%
PM10						
%EBIT	0.44%	0.47%	0.49%	1.07%	0.00%	0.41%
upper	1.28%	1.37%	1.43%	3.11%	0.00%	1.20%
%EBT lower	0.46%	0.65%	0.85%	1.87%	0.00%	0.57%
upper	1.34%	1.88%	2.47%	5.43%	0.00%	1.65%
NMVOC						
%EBIT	0.00%	0.00%	0.09%	0.19%	0.00%	0.02%
upper	0.00%	0.00%	0.27%	0.59%	0.00%	0.05%
%EBT lower	0.00%	0.00%	0.15%	0.33%	0.00%	0.02%
upper	0.00%	0.00%	0.47%	1.04%	0.00%	0.07%
Total externalities						
%EBIT	1.82%	0.87%	0.96%	2.32%	0.57%	1.28%
upper	5.15%	2.37%	2.64%	6.25%	1.55%	3.56%
%EBT lower	1.90%	1.20%	1.67%	4.05%	2.02%	1.75%
upper	5.39%	3.25%	4.59%	10.92%	5.48%	4.87%

3.4 Discussion of the results

According to the 2004 emission data, the externalities of the Greek steel industry, due to the SO_2, PM10, NO_x and NMVOCs emissions range between 1.27 and 3.53 million €. About 40% of this external cost is attributed to the SO_2 emissions of SOVEL. The latter company is responsible for almost 60% of the sector's externalities.

With the exceptions of SOVEL and Halyvourgiki (the latter did not report PM10 emissions probably because they did not exceed the threshold of the EPER Decision), PM10 seems to be the most important source of externalities for the steel companies, followed by NOx. NMVOCs are insignificant from this point of view. Yet, it is mentioned that the external costs of VOCs estimated by the CAFE programme involve serious omissions, because of the failure to account for organic aerosols as well as for impacts associated with long-term (chronic) exposure to ozone.

As far as the 'externality' ratios are concerned, the total environmental costs are estimated between 1.28% and 3.56% of the sector's EBIT and between

1.75% and 4.87% of its EBT. Hellenic Halyvourgia presents the worst performance, since the company's externalities range between 2.32% and 6.25% of EBIT and between 4.05% and 10.92% of EBT. These figures are almost two times higher than the overall ratios of the sector. On the opposite side, SIDENOR is the most resilient company given that the external costs vary between 0.87% and 2.37% of EBIT and 1.20% and 3.25% of EBT. The results of the other companies lie in the middle.

4 Concluding remarks

From the analysis presented, it becomes evident that determining whether implementing BAT in a sector is 'economically viable' under the IPPC framework is not an easy task due to the diversity of the industrial units involved. Considering the Greek steel sector, installation of environmental systems in order to achieve practically zero air emissions would imply different financial burdens to the steel companies under investigation. Hence, an in-depth analysis would be probably more appropriate.

The approach presented offers certain advantages and it could prove to be beneficial in different levels of decision-making process. It recognizes the most significant environmental stressors and it highlights the most 'vulnerable' firms to changes in operating and capital expenses associated with implementing BAT. For example, in the case presented it was indicated that the external cost of the sector was mainly caused by SO_2 and PM10 emissions. Hence, process integrated measures or end-of-pipe techniques should focus on those pollutants. In addition, the 'externality' ratios provided important information, concerning firms' vulnerability. For instance, externalities in the case of Hellenic Halyvourgia were estimated to be up to 10.92% of EBT, which are significant, considering that the profit margin (i.e. the net income as a percentage of the revenue) of the company in 2004 was 1.3%. Hence, at a hypothetical situation in which a new environmental law, requiring elimination of air emissions, was enforced, the viability of the firm would be probably undermined, although only a time-series analysis could provide a more secure answer.

Concluding, comparing externalities with costs for mitigating environmental impacts may be promising but not sufficiently clear due to the uncertainties involved. Yet, it is evident that there is a growing attempt in EU to more systematically incorporate monetary values in private and public decision-making, i.e. in permitting procedures under IPPC jurisdiction, and towards this direction industries must fully understand and implement environmental valuation processes in order to establish a better relationship with the State authorities and the local communities.

Acknowledgements

This study has been undertaken by the GEVAD Project, which is co-funded by the European Social Fund (75%) and National Resources (25%) – Operational

Programme for Educational and Vocational Training (EPEAEK II) – PYTHAGORAS.

References

[1] International Iron and Steel Institute (IISI), *Steel Statistical Yearbook 2007*, IISI Committee on Economic Studies: Brussels, pp. 10 – 12, 2007.
[2] European Parliament, Steel Industry, European Parliament Fact Sheets, http://www.europarl.europa.eu/facts/4_7_2_en.htm.
[3] EUROFER, *The European Steel Industry and Climate Change*, EUROFER: 2000.
[4] Beguin, J-M., *Industrial relations in the steel industry*, http://www.eurofound.europa.eu/eiro/2004/12/study/tn0412101s.htm.
[5] European Commission, *Council Directive 96/61/EC of 24 September 1996 concerning integrated pollution prevention and control*, Official Journal L 257, pp. 0026 – 0040, 1996.
[6] European Commission, *Integrated Pollution Prevention and Control Reference Document on Economics and Cross-Media Effects*, DG JRC, European IPPC Bureau, 2006.
[7] Halyvourgiki S.A., http://www.halyvourgiki.com/english/index.html.
[8] SIDENOR S.A, http://www.sidenor.gr/home.aspx?lang=EN.
[9] Helenic Halyvourgia, http://www.hlv.gr/company-net-1-en.html.
[10] European Commission, *Commission Decision of 17 July 2000 on the implementation of a European pollutant emission register (EPER) according to Article 15 of Council Directive 96/61/EC concerning integrated pollution prevention and control (IPPC)*, Official Journal of the European Communities, L192, pp. 36 – 43, 2000.
[11] AEA Technology Environment, *Damages per tonne emission of* $PM_{2.5}$, NH_3, SO_2, *NOx and VOCs from each EU25 Member State (excluding Cyprus) and surrounding seas*, Service Contract for carrying out cost-benefit analysis of air quality related issues, in particular in the CAFE programme, 2005.

Environmental and climate risks in financial analysis

M. Onischka
Wuppertal Institute for Environment, Energy and Climate, Germany

Abstract

The assessment and consideration of risks in the analysis of companies and projects play an important role in investment decisions. In the context of climate change new significant types of risk arise, such as physical risk, regulatory risk or risk of reputation. It can be shown that climate risks are additive, increasing the whole risk exposure significantly. Relevant studies have shown that depending on the sector climate risks will affect the value of assets (value at risk) up to double-digit percentage. Applying common methods and procedures these risks are generally determined and measured insufficiently. Thus, the provision for environmental and climate risks in the granting of credits, sell-side/buy-side research or due diligences is not adequate, although existing approaches would generally allow a valuation of these risks.

The main target of this paper is firstly to analyse the problem of the 'debasement' of historical data especially in the light of climate change. In addition to that, relevant approaches for the process of financial analysis will be outlined in order to implement environmental and climate risks by means of adjusted risk premiums. In this context these approaches represent practical potentialities for the implementation of risk premiums as well as fundamental factors that have an impact on the investment decision. Here an important connection between risk measure, economic reference parameter and valuation method is developed.

Keywords: climate change, risk assessment, valuation, risk management, risk measure, financial analysis, environmental risk.

WIT Transactions on Ecology and the Environment, Vol 108, © 2008 WIT Press
www.witpress.com, ISSN 1743-3541 (on-line)
doi:10.2495/EEIA080081

1 Background

In practise the valuation of companies – be it the assessment of the 'fair' company value or just its creditworthiness – is usually realized by discounting future profits/free cash flows, sometimes even by means of a comparison of similar companies of a peer group. In any case, only those aspects are considered that influence the value drivers of the business success directly; that are monetary variables (Hitchner [5]). The human influence to the ecological system usually occurs as external effects only. At first they do not have any influence to the monetary or business sphere and therefore are not considered in valuation processes. Concerning normal valuations there is no interest in the objective or true value of a company or project that regards externalities too. On the contrary the market value is relevant only because the valuation is usually made for a specific purpose (e.g. acquisition, stock analysis).

From a macroeconomical point of view such an microeconomical optimization is not welfare-optimal because not considered damages and cost will lead to a more intense use and pollution of the environment (Pigou [13]). By contrast, climate change influences the financial performance of companies/projects directly because in part external effects will be internalized. Climate change has the specific feature that due to the greenhouse gas emissions such a strong feedback by the ecosystem occurs that impacts to the economics system reach relevant financial dimensions. The fact that both politics and society are already reacting – e.g. via political procedures - influences the value drivers of companies directly. Therefore climate change becomes relevant for the valuation of companies and projects, too. In the last years several studies have been published which carefully estimate costs and decline in company values as a result of climate change. The estimations refer to different economic levels (worldwide, national wide, industry-specific) and include a number of simplifying assumptions (for more details and an overview of selected estimations of economical impacts refer to DIW [1], Onischka [14], Stern [19] and WestLB [20]).

Climate change does not only affect companies by extreme weather events like storms, droughts or floods. For European companies future damages and losses caused by weather will turn out to be less significant compared to companies in regions with relative high climate exposure (IPCC [9]). The business development is affected by climate change through three channels: operating business, investments and capital costs. The operating business is especially influenced by changing consumer demand for less greenhouse gas intensive technologies or products. As a result, both cost structure and sales volume are changing: Turnover, cash flows, profit etc. are inevitably affected. The second channel are investments, with long-term investments in fixed assets, production technologies and R&D as the most relevant. Last not least also costs of financing add up from cost of debt and cost of equity (Stern [19]). As soon as banks start implementing aspects of carbon intensity into their corporate credit rating the cost of debt will change (Onischka and Orbach [11]). An increasing part of the investors are also trying to consider the impact of climate change to the business development of their assets. A growing demand for risk-adjusted premiums on the return of equity will be the result.(for instance see: The Carbon Disclosure Project. www.cdproject.net)

If – as a result of climate change – relevant parameters and value drivers change and are predictable under uncertainty only, business-specific risks will emerge (Onischka [14]). In the previous years a distinction into four risk categories has been established (physical risk, regulatory risk, reputational risks, litigational risk)(For a more detailed overview of climate risks, refer to [19]) and influence the value drivers and therefore a company's performance directly. The impression could arise that these risks are already covered by conventional business risks, which are typically estimated in company valuations or ratings. However, climate change has the specific feature that it virtually influences industries and companies as an external effect and therefore does increase/diminish the risk exposure additional to the conventional business risks. The estimates of various up to now works show quite clearly that climate change is economically no zero-sum game, in truth it will bring net losses whereas industries will be affected differently (Stern [19]). The occurrance of effects of diversification may be limited only. In the result economic risks due to climate change are to be understood as additional risk which are covered with the conventional financial analysis/rating only insufficiently.

In this paper the thesis is supported that climate risks are considered insufficiently in conventional methods and approaches. In this context the important issue arises in which shape data and information need to be provided to ensure their usability. Above all, potential differences between on the one hand feasible statistical risk measures and on the other hand relevant risk measures might be interesting, as well as practice-relevant combinations of reference parameters and valuations methods. Other important issues like the practical methods for ascertaining the required data and risk measures (e.g. by means of data simulation or subjective probabilities) is not covered by this paper. The structure of the paper is as follows: Firstly, in chapter two the attention will directed to the analysis why historical data seem to be not feasible for the consideration of climate change as well as other environmental factors and therefore new approaches are required in future. In chapter three a simple systematic will be worked out that shows how climate risks shall be measured resp. quantified to ensure their ability for valuation in practice relevant methods. At this a distinction between risk measure, reference parameter and practical approach will made, leading to an assessment that will carry out useful and feasible combination of factors of each category.

2 Debasement of historical data

In the centre of company/project assessments stands the analysis of the current situation of a company's assets, liquidity and profitability. Based on this status quo as well as historical data of the company, trends are derived and projections are calculated. Typically, selected key business ratios are estimated to facilitate forecasts for future periods. For simplification reasons trends or the status quo are often extrapolated. So in general historical data are the starting point for any assessment or valuation. Concerning the concrete methodology different approaches are used. Sometimes only the asset value, sometimes only the capitalised earnings value, sometimes comparisons based on key business ratios, and sometimes hybrid forms are used. The methodical alternative at project/company valuations is the summation of future costs as well as profits. Experience and cost data of the past are usually the fundament of any calculation, too.

Regarding climate risk there are several specifical requirements on data quality. Existing historical data of weather or weather extremes (e.g. quantity and power of storms) are ineligibly for forecast (Schwartz and Randall [17]). Appropriate climate models show that there was, there is and there will be non-linear, time-delayed and even erratic changes at direct and indirect climate effects (Fischer et al. [2]). Therefore it is problematic to draw conclusion from the historical climate exposure (that is the monetary priced impacts of risk as a result of climate phenomenon) because this will result in significant false estimations. This is even true although a multitude of sectors (e.g. agriculture, fishing, logistics/ transporting, tourism) have already been exposed to physical climate and weather risk like heat/cold waves, high/low water, storms or El Niño. But for the available economic data (ca. 200 years) there were no fundamental changes of climate in such a scale, which would have had significant impacts on cash flows of companies. So it can be stated that without any adjustments the empirical climate exposure has no direct significance, the historical data are so to speak 'debased'.

However, physical data are not the only information that has influence on the business success. Especially future regulations and changing reputation play a central role. Empirically, regulations can be estimated by logit- and probit-models.(for an introduction to such models refer to Ronning [15]) But these models presume that the structural connection between impact factors and regulations will last for the future. But for climate-related regulations various problems remain. For example there are – if at all – only a few relevant and comparable regulations. Observable events only occur within large intervals (e.g. in terms of legislation) and their frequencies make strong statistical conclusions almost impossible. As soon as such events occur they will have significant impact on business developments. As a rule, a certain regulation (e.g. limits for CO2-emission of car per kilometre) is related to a certain object and appears in this form only once. Trends or causalities based on a historical development are not possible because of missing comparability. So it can be emphasized again that historical data – be it data of weather related events of regulations – are unsuitable for the assessment of future development. However, this result contradicts the already discussed and established practise to extrapolate the historical development to the futures with small adjustments only. The question remains what are adequately solutions to avoid wrong conclusions as a result of wrong underlying database. Here are three possible suggestions:

Possibility 1: Lump-sum risk premiums are chosen in an extent that they cover possible divergences of the conventional estimated variables in an acceptable confidence interval. Though, this would have the disadvantage that due to the ignorance about future events, risk premiums turn out to be considerably bigger than actually required. The result would be a systematic undervaluation.

Possibility 2: For several time frames estimations about the net cost of climate change already exist for national and industry level (ref. to chapter 2.1). In the context of a Monte-Carlo-simulation qualities of historical data series are connected with such kind of additional parameters to new synthetic, simulated data series.

Possibility 3: Subjective apriori probabilities will be referred to several formed scenarios and finally used for the calculation of a company's value. Although these probabilities can be derived from just from the analyst's a priori probabilities; however, the Bayesian Risk Management usually uses an aggregation of expert based knowledge. By this approach existing information about climate change or its consequence that is not included in formal data can there be made usable up to risk measures.

Due to the multidimensional influence of climate change there is none state of art approach that enables to deal with this data-relating problems in practice. The approaches, which are on hand or still to be developed (e.g. Bayesian Risk Management) are not yet tested in practice in a scale that allows characterizing a satisfactory solution of the problem (Haas and Jaeger [4]).

3 Relevant approaches in financial analysis and risk controlling

3.1 Interrelationship between important elements in risk management

At the valuation of companies or projects risk itself plays a subordinate role only. Rather the focus is on a preferably realistic, monetary valuation of the object (Spreman [18]). In this meaning risk has merely to be understood as a probability for a possible variation of the estimate from the 'real' value on a certain scale. This overall risk arises primarily in consequence of the basic assumptions of the assessment. On the one hand, these are the model assumptions. However, on the other hand the estimates and forecasts about the future financial development do have a stronger effect. If the risk shall be considered in the valuation in a way that also the target value (project or company value) could be adjusted, the risks must be treated systematically within the whole valuation process. These (risk) valuations are not only used for financial analysis; risk analysis and assessments of all sorts of valuation objects take place at financial service providers in the context of risk controlling particularly. In this context risk controlling has to be understood as part of a general, often interdivisional risk management, which identifies risks and measures financial consequences. In addition it is also used for the control of the risk policy (Johanning et al. [8]).

In practice, information about climate-related risks appear in two forms: First as an uncertainty about value drivers of the financial figures of companies;

second, as more general external information, for instance scientific or expert-based estimations about the net cost of climate change for economies or selected industries. To ensure a systematic consideration of both forms of information they should have some influence on so-called risk measures. At this, risk measures are the mathematical description of the characteristics of these risk information. Usually these kinds of information are extracted out of statistics or stochastic, sometimes only rough risk information is accessible. For example in the recent reports of the Intergovernmental Panel on Climate Change (IPCC) probabilities to certain statements and results are only made verbal.(certain verbal expressions like "very likely" will be referred to certain range of probability. In the case of "very likely" this verbal expression is often referred to a probability between 90% and 99%.)

The valuation process before the use of any valuation method is similar for both financial analysis and risk controlling, especially regarding its content and procedure. Therefore the following analysis covers both areas. Figure 1 illustrates a simplified connection between the elements of the valuation process. Both in practice and in literature a clear separation between these elements is not made. After their collection, the 'financial data' are condensed and as a rule important variables (for example in form of value drivers) are identified that might have a big influence to the company's value resp. to the valuation item. This step is called later 'data analysis'. Either risk information refers to these data directly or/and external risk information will be added. External risk information in this context means that general information about risk and probabilities – e.g. information from IPCC or estimations of relevant studies about certain industries – will be taken into account. As a general rule all information will be condensed to a single 'risk measure'. Since the aim of the valuation process is a risk adjusted valuated variable, the risk measure is used along with an economical 'reference parameter'. Here, the risk measure indicates the ‚amount' of risk meanwhile the reference parameter can be seen as a ‚unit' of an economical parameter. It is quite obvious that the characteristics of the risk measure depend on the kind of reference parameter. One example: Imagine a company with a high carbon and climate risk exposure. The risk information will highly differ if they are referred to the company's profitability, company value or its market share. To some extent the methods of valuation discern significantly, for technical reasons the use of just one general reference parameter is therefore not feasible.(this might be one of the main reasons why there does not exist only one single standard method for the consideration of climate and environmental risk.) If this context is taken into account, each in practice relevant method will entail one (or few) economical reference parameter; this reference parameter will entail a certain group of risk measures. In contrast to conventional subjects of risk assessment the effort to achieve these risk information can be significant higher and more time-consuming. Therefore, regarding climate change it is crucial to know about which variable risk information is needed and which kind of measure fore this information must be provided. The discussed simplified interrelationships between these elements are visualized in figure 1. In the following chapter the mentioned elements of the valuation process – i.e. risk measure, reference parameter and valuation method – will be discussed separately and then consolidated with regard to practicable combinations.

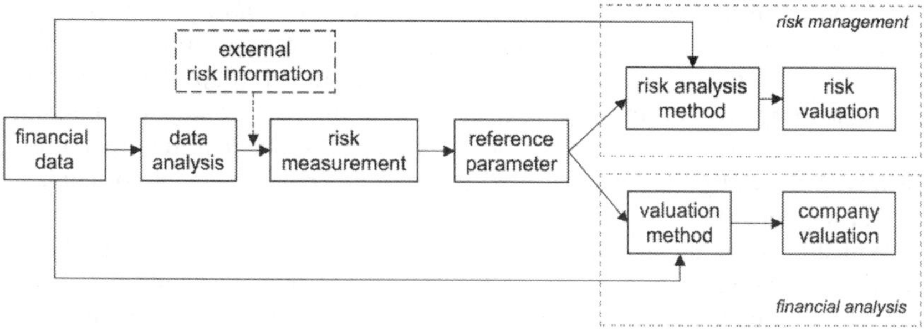

Figure 1: Interrelationship of important elements within the valuation process.

3.2 Systemization of risk measurements, economic reference parameters and methods for valuations

Risk measures describe identified risks quantitatively by suitable probability distributions. In practice, usually a separate estimation of the expected amount of a payment and the risk of its settlement is carried out. Risk measures can refer to single risks (e.g. damage of fixed assets) but also to the complete risk of a company (e.g. profit). Risks are often described by either the amount of damage and the probability of its occurrence or by a standard deviation. These and other important risk measures can be systemized within the three following groups. (In general different systematizations are possible as well (e.g. security related vs. failure related) but in the context of this analysis not appropriate. For further systemization of risk measures refer to Reichling et al [14] and Gleisner [3].)

Two-sided risk measures	One-sided risk measures	Other risk measures
Standard deviation	Value at risk (Var)	Volatility (financial)*
Variance	Conditional value at risk (Cvar)	
Covariance/correlation-coefficient	Equity requirement (ER)	
Density function (not normal distributed)	Lower Partial Moments (LPM)	

Figure 2: Systemization of risk measures. *In opposite to the mathematical definition (standard deviation of a random variable) this volatility refers to the definition used in the financial world. In these terms volatility is to be understood as standard deviation of historical changes of an observable parameter.

The choice of an economical reference parameter depends strongly on the aims and intentions of the ‚risk analyst'. While the investor is primarily interested in potential changes on the return of investments, a fundamental oriented analyst is rather concerned about variances of value drivers of the free cash flow of a company (e.g. certain cost structures). Economics reference parameters for risk information can be differentiated into several groups. On the one hand there are relative parameters, like types of '*return of investments*' as well as '*discount factors*'. On the other hand '*value drivers*' and absolute to the type of '*absolute monetary variables*' can be subsumed to absolute parameters. If the assumptions of the standard models in finance would be valid, these groups should provide similar outcomes as a result of their formal relationship. However, in reality the financial markets are neither full efficient nor complete, so that segmentation into these four groups is useful.

Which concrete method is used for valuation strongly depends on the object to be valuated. As a rule in financial analysis this is the value of a company – be it the market value or the fundamental value. In risk controlling however, merely the monetary risk is isolated. So in practice, totally different methods are used. In the Appendix methods for the valuation of companies/projects are listed according to their relevance for practice, as well as approaches used in risk management.

3.3 Synopsis

As already mentioned, the most valuation methods can be assigned to only one (in some case to two) economic reference parameter with whose value risk can be expanded into the valuation. For the methods with a high relevance for practice these matches are listed in the following table. Technically this table is a synopsis of chapter 3.2 and figures 3 and 4 of the Appendix, whereas only methods with high practice relevance are considered. Individual explanations for these matches can be set aside because they are the logical result of the

Practice relevant method for the management of risk	**Economical reference parameter**	**From a logical point of view usable risk measure**
Discounted Cash Flow Equity/Entity *	Discount rate/discount factor	One-sided risk measure
Gross rental method	Absolute monetary variables	One-sided and tow-sided risk measures
Combined approaches **	Absolute monetary variables	One-sided and tow-sided risk measures
Multiplier method	Absolute monetary variables	One-sided and tow-sided risk measures
Capital Asset Pricing Model	Return of investment	Two-sided risk measures
Sharp Ratio	Return of investment	Other risk measures
Sensitivity analysis	Independent from only one economical reference parameter/risk measure	
Risk simulation with VAR variables	Monetary variables	Two-sided risk measures

*: *value drivers may be an alternative reference parameter.*
**: *discount rates/discount factors may be an alternative parameter.*

Figure 3: Combination of valuation method, reference parameter and risk measure.

Method	Main idea	Relevance for practice
Discounting methods		
Discounted Cash Flow – Equity	The net present value of the estimated future cash flows (usually free cash flows) or profits (dividends or annual net profits) will be calculated.	High
Discounted Cash Flow - Entity		High
Adjusted Present Value		Medium
Dividend Discount		Low
Gross rental method		High
Weighted Cost of Capital		Low
Net asset value methods		
Net asset value approach	Assessment of the net value of the assets of a company	Low
Combined approach		High
Liquidations approach		Low
Multiplier methods		
Multiplier method	Deduction of the company value from real purchase prices resp. market values of similar companies	High
Approaches of the portfolio theory		
Capital Asset Pricing Model	Minimization of the unsystematic risk. Valuations are made out of risk-return-combinations of the market.	High
Arbitrage Pricing Model		Low
Sharp Ratio		High
Real options		
Real options	Valuation of possible strategic actions	Low

Figure 4: Systemization of valuation methods (for a more detailed description of these several methods refer for instance to [18] or Hockmann and Thießen [6]).

systematic of these methods. Only one simple example: The basic model of the CAPM makes a direct relationship of the return of a portfolio/security on the one hand, and its risk on the other hand. In this context the central economical reference parameter is the return on investment. Also determined by the methods resp. its assumptions, only one-sided risk measures can be used – in this case variances and covariances. Similar considerations are possible for the other methods. At the end only one feasible combination of valuation method, economic reference parameter and risk measure is reasonable:

4 Conclusion

There is at least one general result of this analysis: Depending on the used valuation method, risk shall only be valuated with economical reference parameters and risk measures that are feasible to the method resp. its assumptions. Though, this also implies that not every kind of mathematical information of risk (risk measure) can be used in practice. This aspect is of highly importance when climate or other environmental risks shall be considered because obtaining this kind of risk information will often take much more effort – for example the mentioned data adjustments in chapter 2 – than the assessment of conventional business risks. The information condensed in the systemization of this paper can be used to get an information what kind of requirements to risk information are needed to ensure an appropriate consideration of environmental risks in used valuation methods. At least the way of the discussed context of this synopsis might provide a good support how climate and environmental risks should be considered in risk assessments. As briefly discussed in chapter two, for valuations of companies/projects several risk categories are relevant but their value cannot be analysed with an historical database only. As soon as appropriate methodical solutions have been developed, regarding modified data can be used in the ways described by the systemization of this paper. Possible outcome might be several practical instruments that are used independently before starting the conventional valuation methods (like discounted cash flow, CAPM or VAR). Hereby climate and/or environmental aspects and risk can be taken into account adequately.

5 Appendix

Approaches for valuations of companies/projects: A variety of methods exist to determine the value of a company resp. project. It is possible to distinguish between five main groups of valuation models mentioned in figure 3. *(The allocation of practice relevance is made out of rankings in literature/European practice guidebooks as well as the frequency of their appearance – even though it remains subjective.)*

Approaches of risk management: Partly, the methods of the financial analysis are also applied at risk management, in particular elements of the portfolio theory. (For instance, based on the CAPM risk adjusted discount rates as well as the complete risk of a portfolio can be determined.) Quite often the result of specific methods in risk management is a discreet characterization of risk in the form of risk groups or rankings. This information will be used directly for activities in minimizing or hedging risks. Many methods of risk management used in practice are based on qualitative heuristics; therefore a reasonable use of quantitative risk information is not possible. For this reason quantitative methods are covered in figure 4 solely. *(The allocation of practice relevance is made out of rankings in literature/practice guidebooks as well as the frequency of their appearance – even though it remains subjective.)*

Method	**Field of application**	**Relevance for practice**
Sensitivity analysis	Financial risks	High
Scenario techniques	Market risks, external developments	Medium
Expectation-value-principle (μ-principle)	Quantifiable risks	Low
Risk simulation with VAR (e.g. ascertainment of the minimum deposited equity according to BASEL II)	Financial risk	High
Lump-sum agios/disagios	Financial risks	Medium

Figure 5: Selected valuation methods of risk management (for a more detailed description of these methods refer for instance to Kuruc [9] or Reichling et al. [14]).

References

[1] DIW, Klimawandel kostet die deutsche Volkswirtschaft Milliarden, *DIW Wochenbericht,* No. 11/2007.

[2] Fischer, H., et al. (eds). *The Climate in Historical Times. Towards a Synthesis of Holocene Proxy Data and Climate Models*. Berlin 2004.

[3] Gleisner, W., Risikomaße und Bewertung – Entscheidungen unter Unsicherheit und Erwartungsnutzentheorie, *Risikomanager*, 06/2006, pp.4–20.

[4] Haas, A., Jaeger, C., Agents, Bayes, and Climatic Risks – A Modular Modelling Approach, *Advances in Geosciences*, No. 4, pp. 3–7, 2005.

[5] Hitchner, J., *Financial Valuation. Application and Models*. New Jersey 2003.

[6] Hockmann, H., Thießen, F., *Investmentbanking*. 2nd Edition. Stuttgart 2007.

[7] IPCC, *The IPCC 4th Assessment Report. Climate Change 2007, Adapation and Vulnerability,* Geneva 2007, www.ipcc.ch.

[8] Johanning, L., Funke, C., Bossert, T., Rendite oder Risiko? Risikokontrolle im Asset Management. *Risikomanager*, 06/2006, pp.14–18.

[9] Kuruc, A., *Financial geometry – a geometric approach to hedging and risk management*. Harlow 2003.

[10] Oehler, A., Unser, M., *Finanzwirtschaftliches Risikomanagement*, Bamberg 2002.

[11] Onischka, M., Orbach, T., Klima und Finanzmarkt. *So investiert die Welt. Globale Trends in der Vermögensanlage*, ed. Bierbaum, D., Wiesbaden 2008. pp. 77-97.

[12] Onischka, M., Climate Change will alter financial markets, *Nikkei Ecology*, No. 8./2007. p.129. [japan.]

[13] Pigou, A., *The Economics of Welfare,* Fourth Edition, London 1932.

[14] Reichling, P., Bietke, D., Henne, A., *Praxishandbuch Risikomanagement und Rating,* Wiesbaden 2007.

[15] Ronning, G., *Mikroökonometrie,* Berlin/Heidelberg 1991.

[16] Shim, J., Siegel, J., *Handbook of Financial Analysis, Forecasting and Modelling,* New Jersey 2001.

[17] Schwartz, P., Randall, D., *An abrupt climate change scenario and its implications for United States national security,* Washington 2003. www.icomm.ca/survival/pentagon climate change.pdf.

[18] Spreman, K., *Valuation*. München 2004.

[19] Stern, N., *The Economics of Climate Change. The Stern Review.* Cambridge 2007.

[20] WestLB Panmure, *From Economics to carbonomics*, London 2003.

Evaluation and land use planning process of a high population growth rate municipality: Los Cabos, Mexico

O. Arizpe[1], J. Fermán[2], R. Rivera[1], J. Ramírez[2] & R. Rodríguez[1]
[1]*Universidad Autónoma de Baja California Sur. La Paz, México*
[2]*Universidad Autónoma de Baja California Ensenada, México*

Abstract

The high productivity, diversity and social-economic importance of coastal systems and their components present a very complex scheme that requires a detailed knowledge of the characteristics of the environment and social-economic factors. This has been designated as high priority in the Municipality of Los Cabos, Mexico, where inadequate development plans, caused by a very high rate of economic and population growth (9% per year), have turned it into an extremely vulnerable area with high pressure from the irregular growth of tourists facilities. The purpose of this study was, precisely, to develop a model of ecological ordinance or land use planning in general for the region. This model could integrate other processes of environmental planning that are currently in operation at municipal level, such as those for regional and marine planning in the Gulf of California. A characterization of all environmental, social and economic components of the municipality was carried out. Eighty one environmental units, or micro regions, were identified and mapped out using a scale of 1:50 000 on the basis of sub river basins, physiography and vegetation, integrated with all the data bases available in a Geographic Information System (GIS) of the whole Municipality. A diagnosis and prospective analysis of different scenarios were also developed, using fragility, pressure and vulnerability indicators. Finally, it was realized that the Los Cabos region needs to change its economic development approach, proposing a general model of ecological ordinance with specific actions and criteria of land use for each microregion.

Keywords: prospective analysis, Los Cabos, land use planning.

WIT Transactions on Ecology and the Environment, Vol 108, © 2008 WIT Press
www.witpress.com, ISSN 1743-3541 (on-line)
doi:10.2495/EEIA080091

1 Introduction

The high productivity, diversity and great socio-economic importance of coastal systems have high potential for economic development. A scientific foundation is needed for adequate management of such areas. A scientific foundation preserves natural resources in an area, identifies new potential resources, and provides the tools to develop suitable management policies for sustainable development of the coastal system as a whole. Ecological Territorial Ordering is a tool to create a diagnosis, based on biophysical and socio-economic variables, arranged in a combination of administrative units (municipalities) and Enviromental Units for improved management (León et al. [1]).

The municipality of Los Cabos is part of the Gulf of California, and is located in the south of the Baja California Peninsula. It is an area of great cultural and historical value, impressive scenery and high tourism potential. All this contributes to the zone´s large capacity to generate important economical benefits, at both regional and national levels. Tourism in Los Cabos is increasing. It is an area of luxury tourist infrastructure, which combined with it´s natural beauty generates a wide range of tourism activities. In recent years, it has become the region with the highest population growth rate in Mexico (9.2% per year, INEGI [2]). This is due primarily to the economic growth generated by the tourism sector, which has resulted in a great increase in the demand for services and infrastructure (CEI [3]) and inmigration.

The main economic potential of the area is located in the housing and recreational sectors.

The problems generated by the accelerated and disorderly growth include conflicts over land ownership, unplanned settlements in the maritime federal areas (especially in the Eastern Cape region), degradation of the coastline and water shortage related problems that have been increasing in the Municipality of Los Cabos and particularly in the Eastern Cape region.

A detailed evaluation is needed of the biophysical and socio-economic characteristics of the municipality, to generate management policies and sustainable development of productive activities, which would improve the quality of life of communities and the conservation of natural resources. This work proposes a model for land use in the municipality of Los Cabos, Mexico, developed from an assessment of environmental characteristics, a socio-economic diagnosis and the suitability of the area. The proposed land use planning would allow a relocation of the land use, and a better planning of productive activities in the region.

2 Methods

Comprehensive and interdisciplinary methodologies, previously used in Mexico to make Ecological Orderings were used in this work. A technical study integrated the development of four stages: characterization, diagnosis, prognosis and proposal. The implementation of each of them has a clear purpose and a series of products, and each step is subject to the guidelines and mechanisms established by Mexican standards.

Characterization The evaluation of the area began with the characterization of the municipality. It identified the study area and integrated environmental information to create zones on a 1/50 000 scale. The zonning established Environmental Units (EU), and define a framework for the analysis of the area from an ecological perspective. It was based on the hierarchical arrangement of three elements that were considered relevant to this process: surface hydrology, topography and vegetation. The resulting map fragmented the municipality into homogenous areas that correspond to unique environmental processes. These are named Enviromental Units.

Diagnosis This was to define which parameters of the established sectors were relevant in the region (tourism, agriculture and conservation) and to develop specific models for each of them. Workshops, involving peoplefrom each sector, were used to adjust proposed ratings in parameter indexes for each sector.

Prospective This phase consisted of calculating the composite indices of Pressure (IP), Fragility (IF), Vulnerability (IV) and the definition of environmental policies for each EU. The IP was composed of two components:

- Index of pressure suitability of the productive sectors and agricultural tourism (PSEC)

PSEC = ITUR + IAP

ITUR = Touristic parameter (Normalizad Value)

IAP = Agricultural parameter (Normalized Value)

- Pressure by Population and Transformed Land Use Index (IPU)
- IPU = IUS + IPOB

IUS = Transformed Land Use Index

IPOB = Population Index

IF was defined as the equivalent index fitness maintenance [4]. The vulnerability index (IV) was the sum of IP and IF. This index spatially defined areas with greater demand for the natural resources of greatest vulnerability.

IV = IP + IF

IP = Pressure Index

IF = Fragility Index (Index of conservation suitability)

Environmental policies were established from the fragility, pressure and vulnerability values. According to Article III of the General Law of Ecological Balance and Environmental Protection (LGEEPA) (SEMARNAT [5]). the relevant policies were:

- Sustainable Use,
- Conservation and Preservation.

Proposal At this stage, the ecological ordering program was developed. It´s an integration of 2 main elements: 1) Model of Ecological Ordering and 2) Environmental Strategies. To achieve this, Environmental Management Units (EMU) were identified through a rezonification of the area from the criteria of the parameters for each sector (the sectorial interactions, [4]). The indices were composed of fragility, vulnerability and pressure, environmental policies by EU and the Natural Protected Areas. Each EMU was assigned an environmental policy on the basis of environmental policies for Environmental Unit obtained at the pronostic stage, according to the proposed categories in Article III of the LGEEPA.

Finally, guidelines and strategies were established for each ecological EMUspecifying authorized, prohibited or conditional land uses for each area.

3 Results

Characterization The most important product of this stage was the zoning, which was a result of integrating the Geographic Information System (GIS), geo-referenced information of the five sub hydrological systems thatare found in the study area, the five types of topography and three main vegetation types present in the municipality. (Table 1). From this integration, 81 environmental units were obtained (Fig. 1), which are compatible with the scale of the work

Table 1: Zonification.

Level	Enviroment	System	
Criteria	Terrestrial	Sub river basins	Key
		Candelaria	3Aa
		Cabo San Lucas	6Aa
		San José	6Ab
		Santiago	6Ac
		Las Palmas	6Ad
Landscape	Unity	Vegetation Type	Key
Physiography	Key		
Dunes	D	Scrub	M
Down hills	B	Deciduous dry forest	S
Valley	V	Forest	B
Hills	L		
Sierras	S		

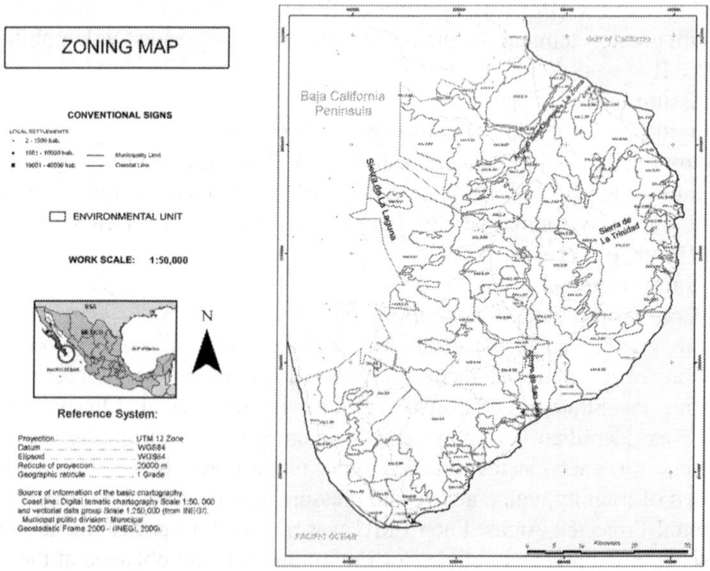

Figure 1: Zoning map.

(1/50 000) and the initial assessment of the suitability of the land, as well as the basis for drawing up the subsequent EMU.

Diagnosis The main outcome of this phase was the analysis of suitability of land for each sector: Conservation, Tourism and Agriculture. The region is very suited to conservation (77% of the area has high or very high suitability), because it´s associated with vulnerable coastal shrubs, lowland forest, wetlands and dunes and unique ecosystems(oak forest, pine-oak forest) that are found largely undisturbed, and species with high conservation value (endemic species and / or those at risk). In terms of the tourism industry, that focuses on traditional tourism, the suitability is greatest along the EU coast (29% of the municipality has high or very high suitability for tourism.). For agriculture, the suitable areas are located along the two main streams and their adjacent areas, largely because of water availability and unconsolidated substrate (57% of the municipality´s surface has high or very high suitability).

Prospective Social and economic dynamics of the municipality of Los Cabos is reflected in the pressure index, which is derived from the analysis of the population, current land use and the suitability for tourism and agriculture. The areas with the highest pressure are the ones adjacent to the coast and the two main streams. The other areas have a lower rate of pressure corresponding to those units with little change of land use and less interest in the productive sectors 44% of the total area of the municipality has a high rate of development and a population growth of 9.2% per annum (INEGI [2]). Given the characteristics of the type of tourism in the area, the greatest pressure from tourism projects are located in the coastal units of the municipality, while the EU without coastline has fewer development projects. 49% of the municipality of Los Cabos has no tourism development proposals while 26% has a high concentration of projects.

The index of fragility, estimated from the index of conservation suitability [4], showed that the most vulnerable areas are the two main streams, the dunes, the Cabo Pulmo EU, and the EU system of oak and pine-oak forest

Calculations of the vulnerability index shows that 50% of the land in the municipality of Los Cabos is at a high or very high level. The EUs with a very high vulnerability are the major streams, the coastal area and the area located to the south of the Sierra de La Laguna Biosphere Reserve, which corresponds to oak forest.This defined the kind of environmental policies set for each of the Eus (Fig. 2)

Environmental policies are presented as follows: sustainable use, 15%, preservation, 17%, conservation, 68%). In general, the strategic scenerio intended for the municipality is of conservation, which is a balance between preserving the natural environment while developing the activities of the productive sectors.

Proposal Integrating all the elements previously developed in the ecological models for thc region, 23 Environmental Management Units (EMU) were composed. Each EMU is a blend of physical, biotic and socioeconomic atributes, and suitabilities. This distinguishes each of them from the rest, with the exception of the EMU for urban-tourism (18) and for the Biosphere Reserve (20) whose rezonificarion for the use of Soil is under the Urban Development Plan (UDP) (Anónimo [6]) and the Protected Natural Area Management Programme (CONANP [7]) respectively.

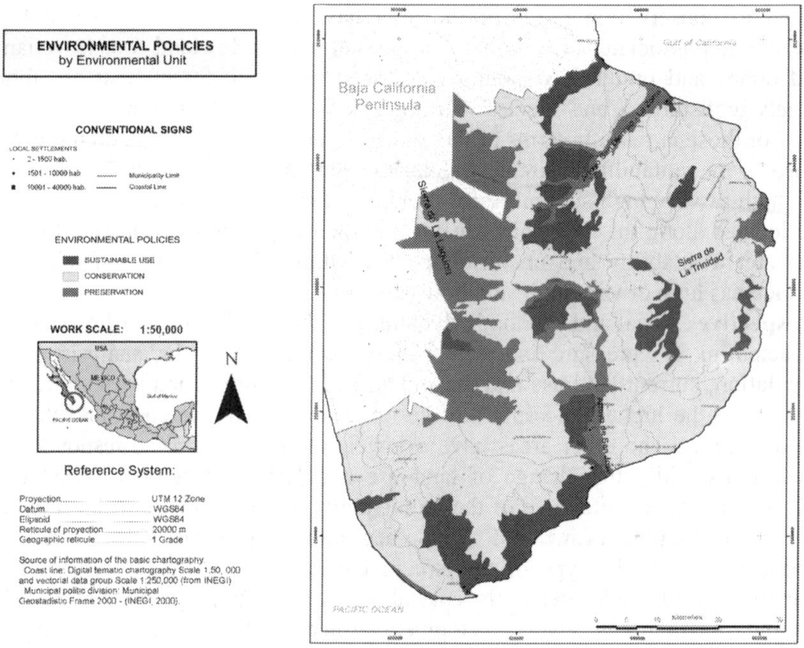

Figure 2: Environmental policies.

Each EMU was assigned one of three environmental policies (Sustainable Use, Conservation or Preservation), that further defined the guidelines reflecting the ecologically desirable state of the EMU, environmental strategies that represent the specific objectives to achieve the ecological plans and a series of specific actions that define strategies for land use for each EMU. Thus the ecological model for the region has three modes that are defined by different policies:

A) Sustainable Use Five EMUs were defined within this policy, corresponding to the EMU 18 (urban-tourism), 19, 21, 22 and 23 that cover 22% of the municipal area (Table 2). Sustainable use promotes the use and management of natural resources in an area and includes the EMU with current or potential use

Table 2: Environmental policies.

Environmental policies	EMU	Area	
		Hectares	Percentage
Preservation	6	80920.66	21.58
Conservation	12	211227.74	56.32
Sustainable use	5	82885.00	22.10

of resources, always respecting the fact that any use of natural resources should consider the principles of sustainability. The ecological guidelines for this policy is to develop sustainable productive activities according to the suitability of the land. The strategies developed to achieve this are: to promote the growth of sustainable farming, diversify productive activities, promote low-impact tourism activities and plan urban growth and tourism.

B) Conservation Twelve EMUs were defined for this policy, EMU 6 to 17. This policy includes 56% of the land (Table 2). This category is regarded as an intermediate environmental policy, between sustainable use and preservation, which was applied to the EMU in which productive activities or urban developments can be carried out while considering the protection of environmental services, provided by natural resources . The ecological guidelines established for this environmental policy is a balance of sustainable development and productive activities, being compatible with the conservation of natural resources. The strategies developed include the identification of natural resources and their conservation status, the study of areas suitable for the development of productive activities under a scheme of sustainability (low-impact alternative tourism, organic agriculture and cattle ranching), the implementation of techniques to reduce the impacts of productive activities on natural resources.

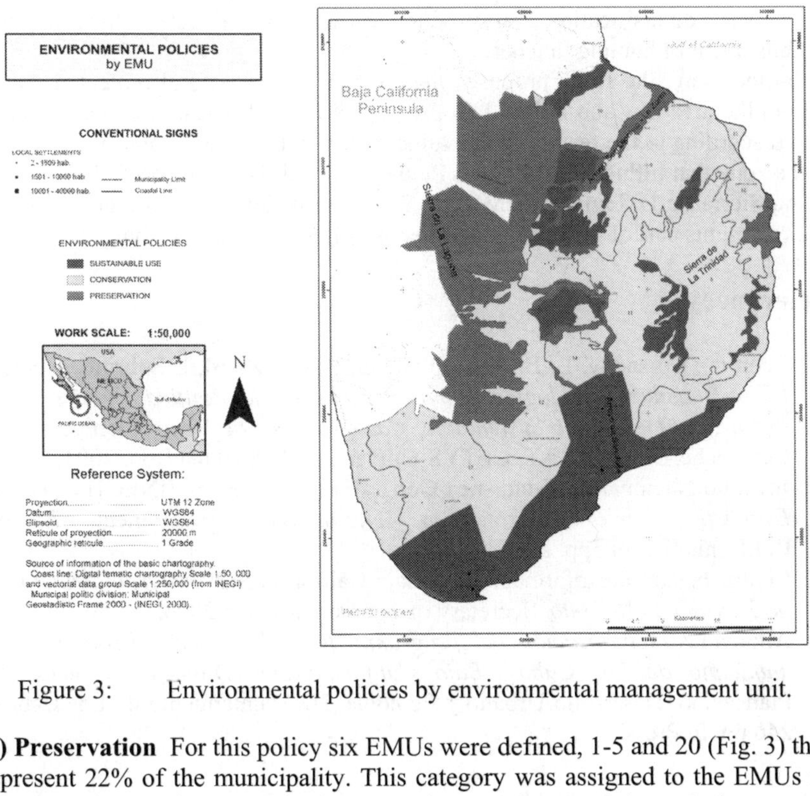

Figure 3: Environmental policies by environmental management unit.

C) Preservation For this policy six EMUs were defined, 1-5 and 20 (Fig. 3) that represent 22% of the municipality. This category was assigned to the EMUs in which it is sought to maintain the relevant characteristics of the natural environments. In addition, activities related to the protection of its biophysical components of the ecosystem are encouraged, while limiting productive activities andhuman settlements. The ecological guidelines defined for this policy is to protect ecosystems and ecological processes.

In summary: The municipality of Los Cabos needs to change its economic development model. The specific actions proposed for the region according to the results of the four stages are:

- The conversion of traditional farming systems into organic ones.
- The change from extensive ranching to a system using fenced areas and supplementary feeding.
- The development of various forms of alternative low-impact tourism.
- Rational water use.
- The research into alternative sources of water supplies for urban, agricultural and tourism use.
- The development of infrastructure away from residential areas at risk.
- An orderly growth of urban and suburban infrastructure outside the EMU urban-tourism.
- The development of low and medium intensity tourism along the coastal strip depending on the characteristics of the area.
- The relocation of existing productive activities.
- The diversification of productive activities directed towards the establishment of hunting ranches.

Furthermore, the study proposes the creation of two natural protected areas: One in the area of Cabo Pulmo-Los Frailes (EMU04) because it is a vulnerable area (according to the results of the vulnerability index) and because it is located in the zone that influences the Cabo Pulmo National Park coral reef. The second in the Sierra de la Trinidad (EMU18), is a zone of interest to the government, and represents part of the low caducifoleous rain forest of the region.

References

[1] León C., I. Espejel, L. Bravo, J. Fermán, B. Graizbord, L Sobrino, J. Sosa. *El ordenamiento ecológico como instrumento de Política Pública para impulsar el Desarrollo Sustentable: Caso Noroeste de México*. University of Campeche, SEMARNAT, CETYS-University of Quintana Roo. 2004.

[2] Instituto Nacional de Estadística Geografía e Informática (INEGI). *Anuario Estadístico.* Baja California Sur. INEGI. State Government of Baja California Sur. 485pp. 2006.

[3] Centro Estatal de Información de Baja California Sur. *Programa Nacional de Turismo 2001–2006*. Secretary of Tourism. México 2006.

[4] Anonymus *Actualización del programa de Ordenamiento Ecológico del municipio de Los Cabos, Baja California Sur*. Dirección General de Planeación, Desarrollo Urbano y Ecología. H. Ayuntamiento de Los Cabos. 246 pp. In Press.

[5] Secretaría de Medio Ambiente y Recursos Naturales (SEMARNAT). *Reglamento de la Ley General del Equilibrio Ecológico y Protección al Ambiente*. Diario. Oficial de la Federación. 2003.
[6] Anónimo. 2005. *Plan Municipal de Desarrollo 2005-2008.* IX H. Ayuntamiento de Los Cabos, Baja California Sur. Septiembre 2005.
[7] Comisión Nacional de Áreas Naturales Protegidas (CONANP). *Programa de manejo de la Reserva de Biosfera "Sierra La Laguna".* CONANP. 209 pp. 2003.

Estimations of costs for dismantling, decommissioning and associated waste management of nuclear facilities, and associated impact on decision processes, functioning of markets and the distribution of responsibilities between generations

S. Lindskog[1], A. Cato[1] & R. Sjöblom[2]
[1]*The Swedish Nuclear Power Inspectorate, Sweden*
[2]*Tekedo AB, Sweden*

Abstract

The future dismantling, decommissioning and associated waste management of nuclear facilities constitute very substantial liabilities worldwide. Legislation and systems are in force in several countries to ensure that funds are available at the time when they are needed and that it is, according to the Polluters Pays Principle, the generations that benefit from the use of the nuclear facilities that also carry the financial burden.

This in turn constitutes a number of challenges which warrant proper attention, and which are dealt with in the present paper: how to carry the burden of financing, the need for securing the funds until the time when they are needed, and the need for precision in the cost calculations with regard to a number of factors. The latter includes the approach selected regarding fairness, in the allocation and distribution of the liabilities between generations, the requirements regarding the functioning of the system of finance (sufficient but not superfluous funding) and the quality warranted for the bases for various decisions needed (e.g. potential investment opportunities in new nuclear research reactors and/or nuclear power plant).

It is described how sufficiently precise cost calculations might be achieved using appropriate calculation methodologies in combination with radiological characterisation, technology selection and financial uncertainty analysis. Examples are given from authentic Nordic work on fuel cycle laboratory and pilot scale facilities.

Keywords: nuclear, decommissioning, decontamination, dismantling, radioactivity, liability, fund, cost calculation, estimate, environmental, legislation.

1 Preamble

It is expected that perhaps hundreds of nuclear power reactors and other major nuclear facilities will be closed during the next few decades and thereby be awaiting dismantling and decommissioning. Such a closed facility typically represents a large negative monetary value with a negative cash-flow – perhaps even comparable in magnitude to that of the initial investment. The main reason for this is the presence of residues of radionuclides, which may well cause the cost for decommissioning to be a couple of orders of magnitude higher than would otherwise be the case.

Historically, responsible action in this regard has varied considerably between different types of nuclear facilities and geographic regions. In many cases, it has been left to Governments to finance remedial actions. A typical environmental project has had to compete with numerous other ones for funding and the result has typically been too little action at a late stage when the problems have already escalated.

There is a growing awareness internationally that appropriate and responsible decommissioning as well as waste management and disposal are fully integral parts of the utilization of nuclear energy. Consequently it is necessary that methods and techniques be developed and applied such that any effects on environment and health are small.

Hence, effective technical tools exist of many kinds for safe and efficient dismantling and decommissioning of various nuclear facilities. However, financial tools are also needed such that sufficient but not superfluous funding is made available at the time when it is needed.

The issues of appropriate and prudent funding as well as estimations of the funding needed are closely interlinked with a number of technical issues, thus making the full analysis of this system complex and difficult. There is a growing awareness of this internationally, e.g. through activities by IAEA, OECD/NEA and EU/COM as well as by many workers in the area. Systems for financing are being established in different countries, and experience is being gained on how to achieve the goal of adequate funding at the appropriate time for efficient and safe decommissioning of the various kinds of nuclear facilities.

The establishment of systems of finance requires that the needs and prerequisites are identified and established, and that difficulties are dealt with, feasible approaches found and effective methodologies developed. It is the purpose of the present paper to provide some detail and examples in this regard. It is largely a result of recent work [1] financed by the Swedish Nuclear Power Inspectorate and also the Nordic Nuclear Safety Research.

2 Needs and prerequisites

The polluter pays principle. The principle that it is the polluter that pays is now relatively generally accepted and established. In concordance, the IFRS (International Financial Reporting Standards) and the IAS (International Accounting Standards) that apply to stock companies in many countries have very strict rules on reliable and precise estimations of liabilities, and secure protection of the corresponding assets including monetary funds.

Identification of the polluter. The polluter is the one that is reaping the benefits of the operation of the facility in question. A corollary to this identification is that no burden should be placed on future generations, e.g. no encroachment should be allowed on the future consumption level due to any remedial actions, now or in the future.

The level of environmental quality after restoration. In theory, the liabilities should include dismantling and decommissioning of older nuclear power plats and other facilities as well as restoration of the land to initial conditions, i.e. *green field conditions*.

However, imposing such requirements within e.g. the EU area would undoubtedly be associated with considerable costs. Consequently, it has been discussed if *brown field conditions* might be acceptable. Such conditions would mean that some radioactive components, e.g. heavy structures, might be left indefinitely.

For the purpose of the present paper, the brown field conditions are discarded on grounds of it leaving responsibility of fairly easily accessible radioactive matter to future generations indefinitely, thus not complying with the polluter pays principle.

The link between the polluter and the restoration. It has also been discussed that nuclear site areas might be used for successive generations of nuclear facilities, and that green field conditions in this perspective might be required only at the end of a period of perhaps three to four nuclear generations. Such an approach might imply that there may up to 100–150 years between the first reaping of benefits and the final decommissioning of the land.

In the application of the polluter pays principle, it is important that the link between the allocation of assets and the full cost for the restoration be sufficiently strong. Extending the time frame over several human generations raises a number of questions including the one of the stability of the society. Moreover, restoration after just one generation is feasible. Consequently, the approach of successive use is also discarded from further consideration in the present paper.

Need for harmonization of requirements. There has been a successful deregulation of the energy market within the EU. This deregulation presupposes that the environmental requirements are equal for all producers, distributors and consumers. Otherwise some energy companies might be tempted to apply minimum or even inadequate environmental standards in order to gain a competitive edge.

Proper priority to liabilities. Business news media report that the median occupation time for managerial positions in industry may be only a few years. Since the environmental liabilities for nuclear facilities extend over decades it would be naïve to assume anything but that long term financing of environmental liabilities are likely to receive a lower priority as compared to e.g. short-term or quarterly profits. This may regard the level of ambition as well as the efforts put into the estimation of future costs.

It was said above that the international rules on accounting imposes stringent requirements on proper priorities. This provides ample opportunities for various environmental and accounting authorities, auditors, investors and shareholders to gain insight and to oversee the management.

Systems of finance. In a number of countries and also within the EU, systems of finance have been established with requirements and oversight of the technical and financial planning processes. In such systems, cost calculations are reviewed by Government Authorities, thus affirming the process through e.g. Government decisions. The prerequisites for such processes include that there exists methodologies for reviews and assessments, independent competence and a knowledge base compiled through independent research. Systems of finance provide insight and assurance for the public, security for the facility owner and information to the tax authorities that the allocation of funds to cover future costs is not actually yet another way to defer or even avoid taxation.

A system like this also has to take into consideration that the financial accounts of the private enterprises that operate nuclear facilities are balanced against the budgets of the Governments that oversee them. It should also be recognized that in some cases, budgets of local municipalities are based heavily on revenues from nuclear power reactors. Thus, all so-called externalities (external effects) should be accounted for and enclosed within the funding system, including the various modes of distribution of responsibility between different stake-holders as well as between generations.

Funds controlled by the Government. It can be discussed whether funds and securities should be located at the industrial companies or be managed directly by a Government organization. The highest level of credibility is probably achieved with funds managed by impartial and competent Government officials. However, a general constraint in that the funds should be completely external and fully segregated from the accounts of companies as well as the regular accounts of the Government.

Quality of the planning process. It might be tempting to assume that the planning process including estimations of costs for decommissioning is similar to that of the erection of any new industrial plant. There are actually huge differences, and for a number of reasons. Moreover, the technical planning is closely interlinked with the cost calculations, since the selections of technologies to be applied depend strongly on their respective costs. The costs, in turn, depend strongly on the radiological situation in combination with various features of the design and operation, some of which might be difficult to identify and evaluate beforehand. Experience as well as documents issued by IAEA and other organizations unanimously show that a very thorough and qualified planning is required in order to achieve the level of precision needed for the requirements on the management of liabilities as well as on the functioning of a system of finance.

Robustness. As the link between the benefits of an operation and the restoration of its facilities and site is being stretched, possibilities may appear for various unplanned events, including accidents. It is highly desirable that a system of finance is robust with regard to such events.

Changes in this regard might be triggered by some exogenous factors and stewardship might be lost. One example of this is the financing of decommissioning and dismantlement of the reactors in Chernobyl after the Ukraine had been separated from the former Soviet Union. Another example is the funding of decommissioning of the six reactors in the power plant "*Bruno Leuschner*" in Greifswald, which was accomplished by the transfer of the responsibility to the financial stronger parts of the Federative German Republic. It is envisaged that further similar changes may appear in the future, thus pinpointing the need to develop systems for adequate funding of future decommissioning.

3 Suitable approaches and methodologies

One major consideration in meeting the needs and considering the prerequisites above is the full realization of the multidisciplinary character of the area. Decommissioning is a discipline of its own in nuclear technology requiring special competence regarding the pertinent approaches in a decommissioning project as well as regarding the methodologies to be applied. For instance, radiological characterization and mapping for the purpose of decommissioning is usually very different from that warranted for day to day operation of the facility in question. It is important that the various skills and competences needed are represented, and that the dictatorship of the majority (usually nuclear engineers) is avoided, or at least balanced or mitigated. Most successful decommissioning projects have had heterogeneous groups with recurrent meetings, thus ensuring adequate responses to upcoming issues as well as propensity and flexibility to change approach when appropriate.

A key issue is the financial planning and review, and the associated combination of financial and technical competence. The ability to understand the prerequisites for the decommissioning of a facility is the key to prudent and appropriate estimates and thereby also to the availability of sufficient, but not superfluous, funding at the time when it is needed.

Focus is often placed on the tools (computer codes) applied for the estimation of the costs as well as on the budgetary classification used. The codes can provide very exact numbers if volumes and lengths of various entities are entered, and using the same budgetary classification among different projects simplifies comparison immensely.

However, uncritical use of such tools might be grossly misleading since various treacherous features may strongly influence the results. In the most extreme case the calculations will only be reproductions of calculations made on a similar facility at another time.

The cost of a project is closely associated with the radiological conditions, and thus codes that enable reliable results include parameters for degree of difficulty. This means that the outcome is not based entirely on per volume empirical parameters but also on difficulty factors that are established through empirical parameters in combination of assessments of the degrees of difficulty for the facility in question.

The result can still not be trusted in general. The experience is that one has to go much further into the various possibilities available for improvement. This as well as actual calculations has been the topic of much of the research carried out by and on commission by the Swedish Nuclear Power Inspectorate, partly within the framework of a collaboration among the Nordic countries Denmark, Finland, Norway and Sweden [1]. Many of the findings have been compiled in an informal guidance document:

The cost estimation should be preceded with the following:

- A radiological survey tailored especially to meet the needs for cost calculations. Such a survey may include e.g. core sampling from a biological shield.
- Sampling design, including use of equipment for measuring
- Methodology selection based on the radiological survey. The selection should include alternative methodologies in case new information is appearing during the work
- Identification of potential cost risers as well as evaluation of the most important ones.

The work should include literature studies and plant visits of similar facilities and projects. Such communication might lead to improvement of the cost calculations through the introduction of actual costs for parts of facilities together with various scaling and weighing factors.

Much of the deviance between estimated costs and outcomes are actually related to various cost raisers. People are frequently anxious to present success stories at international conferences, and indeed, much can be learned from good examples. However, there are also a lot of lessons learned, but they are not as frequently reported on in the literature and on conferences. Therefore, in order to get input for proper planning and appropriate cost calculations, it is necessary to network with people in other facilities and learn from their experience.

It has been estimated in the above mentioned Nordic co-operation that by using the above approach and in reasonably uncomplicated cases, a precision in the cost estimate of ±15% might be attained even for a research facility and at early stages of planning. It is notoriously difficult to estimate the costs for such facilities since they are built for a diversity of purposes, are frequently one of a kind, etc. The figure quoted should be used with great caution since there are numerous examples of much larger deviances between calculated and incurred costs.

The Nordic project has also included working through specific examples of old research facilities, one from each participating country. Although the countries are small, there is quite an abundance of old research and development facilities since these countries had very ambitious programmes for development of nuclear technology and nuclear power generation from the early fifties and at least a couple of decades onwards.

4 Conclusions

Our studies – with special emphasis on the Nordic co-operative project mentioned above – have shown that there are systems available for capital budgeting for future costs. They can be used as platforms for including all costs and benefits in a Cost & Benefit Analysis given the total and overall assets and liabilities of specific nuclear facilities. The demands from the society – even at an early stage – for comprehensive and complete estimations of the future costs for the residues from nuclear activities has necessitated the establishment of the current system, the scope of which can be widened to include essentially all costs to society from generation of electricity by means of nuclear power.

These systems can with fairy straightforward measures become fully integrated and compatible with the legislative demands on private enterprises regarding protection of assets to cover all liabilities for future environmental remediations as required in the IFRS (International Financial Reporting Standards) and the IAS (International Accounting Standards). Such a system requires that costs be estimated in a reliable way, which in turn presupposes development and use of good practice. The following features are essential and obtainable in this regard:

- Better estimates at an early stage of the expected live span of each individual site
- Better financial systems where funds are pinpointed for each facility/site.
- Clearer and non-ambiguous rules for free release and alternative use of land
- Robust cost calculations such that any myopia of the present generation cannot give rise to any costs to the future generations.
- Development of methodologies for evaluation of environmental liabilities (including descriptions, demonstrations and calculations) in European Union and other international co-operation.

Reference

[1] Lindskog, S. et al. Summary of some recent work on financial planning for decommissioning of nuclear research facilities. To be published as SKI Report 2008:xx and available at www.ski.se.

Visualization service for grid-oriented applications of natural disasters

E. Pajorova, L. Hluchy & L. Halada
Institute of Informatics, Slovak Academy of Sciences, Slovakia

Abstract

Every year, forest fires cause enormous damage to vegetation and fauna, environment and property and bind significant human resources. Particularly in national parks and nature reservations, unique areas of high degree of protection can be devastated by fire. For instance, during the destructive forest fire in the Slovak Paradise National Park (Slovakia) in 1976, very unique vegetation was destroyed in the Kyseľ Gorge, where the recovery into the former state will take 200 years [1]. Until now (thirty years after the fire), this locality is closed for tourists because of the vast damage. The topic of a lot of projects is how to prevent such disasters. Our research institute is oriented on GRID computing. A lot of international projects oriented on natural disasters utilise grid computing and within grid solution raises requirement of visualization service for presentation of the intermediate or final results. The basic aim of our research is to create a visual service for modelling and 3D rendering of natural disasters, before fire, flood and landslides. Changing the input data of fires spread used to generate new outputs very quickly. Grid computing on a lot of Clusters and 3D visualisation service can allow new scenes of fire spread. Outputs are used for far adjustment to liquidation fires or floods and landslides.

Keywords: forest fire, GRID applications, natural disasters, visualization service.

1 Introduction

3D visualization service for animation of natural disasters should integrate visualization requests of any kind of application solved in our institute and before solved in international projects oriented on environmental problems. The natural disasters like fires and floods become subject of science in research

WIT Transactions on Ecology and the Environment, Vol 108,
www.witpress.com, ISSN 1743-3541 (on-line)
doi:10.2495/EEIA080111

institutions more and more often. Many applications from this area are using different kinds of simulation tools, which are producing output data for displaying the results of the computation. The purpose of the visualize service is to model and display results of various simulations of natural disasters such as fire spread in time, fire intensity, flood velocity, landslide activity etc. Such service requires unified standards like integration of input data formats and especially creation of unified visualization tool.

This paper is describing such a visualization service developed and tested in our institute, create a 3D visualization service for GRID oriented natural disasters applications. The purpose of 3D viz. service is to model and display intermediate or final results of various simulations of natural disasters like fire spread in time, its intensity and erosion or floods in time or landslides as well. The output of the service is various scenes of terrain by different simulation outputs. Output of the service can also be the files representing the virtual reality of natural disaster and also files, which are generated as input for VR-Systems. 3D service was tested with outputs from applications, which were solved in our institute and also by data from applications of the MEDIGRID project [13].

Particular demonstrations in this article are from fires in Marseille industrial port in the southeast. The second is from a large fire in the Krompla region [1], which is part of the Slovensky Raj. Flood demonstrations are from Povazie, the region around the river Vah. The root of visual service is to create a modelling tool. A modelling tool consists of modules, which were necessary to create. Each module is a UNIX shell script in which is prepared the start of the executables. The modules are divided into three groups according to what kind of output 3D models the group is generating. The functionality of each group is described by appropriate schema.

2 Generating 3D models

- Models of terrain or model of environment.
- Models of simulations.
- Virtual reality models of terrain and simulation (fire spread, flood, landslide etc.).

2.1 Models of terrain or model of environment

The first group of modules developed by us in the 3D viz. service is used for the creation of 3D terrain models and environment models. For 3D models of this type the following modules have been created: CONVERTOR, TERRAIN, IVFIX and CONVERTTOWRL. Examples: (see Figs. 1 and 2).

Advances in computers and information technologies in the last decades stimulate the development of various useful program systems for Natural Disasters (ND) fighting. Particularly, disasters behavior predicting systems can be directly used for specific purposes of fire management, or they can be included in more complex decision support systems (e.g. pre-suppression planning systems). For example for FIRE - behavior predicting systems are

Figure 1: Krompla region – terrain model.

Figure 2: Marseille industrial port – terrain model.

capable to simulate the forest fire front growth after the fire detection [9]. They describe not only the spatial and temporal behavior of forest fires (fire spreading rate and direction), but can quantify and often even display various fire characteristics (e.g. fire intensity, flame length, etc.), which can be useful for the purposes of fire effects analysis. They can be used for simulation of various fire scenarios in a certain region under different conditions to test the fire management response for the fire event (prevention). Most suppression decision support systems are based on fire behavior prediction and make it possible to test the effectiveness of different types of suppression strategies and tactics, taking into account the existing fire fighting infrastructure and specific conditions that affect the fire fighting (e.g. location of water sources, Fire Fighting Headquarters, road network, etc.). The fire behavior models are also used for the reconstruction of previous forest fire events (post-suppression). Such simulations are particularly useful to understand the circumstances that lead to human incidents due to fire fighting [4–6]. Fire behavior predicting systems can also be useful for operational training. Currently, in general, little use of the forest fire decision support systems in operational fire management in the European Union is a great challenge for scientific and development activities within current fire research [2]. Although a forest fire hazard in Slovakia is relatively small compared to other forest land countries, the fires do occur and they are feared for their potential resulting damages. Moreover, during the forest fire in the Slovak Paradise National Park in 2000, the fire had threatened people's lives and nature reserve in the national park. During this forest fire there were various opinions on how to fight the fire in relation to protection of nature and ground attack tactics available. The speed and efficiency of the fire suppression were also very important. Unfortunately, there was no fire simulation model available for the support of strategic fire fighting decisions. On the basis of this experience, the use of a forest fire simulator appeared to be reasonable. From this reason, some national institutions stimulated the investigation of computer-aided forest fire growth simulation and visualization.

From the existing systems available, the FARSITE fire simulator was used for simulation of forest fires. The tests performed confirmed that this simulation system could be used under Central European conditions [8] (comp. also [3]). However, there are some factors, which complicate this effort. Primarily, the topography of the terrain can be very complicated and some phenomena (for instance the so-called chimney effect) could be described to a certain extent only. Moreover, the standard NFFL [7] fuels are not adapted to Central European conditions, therefore new fuel models for those territories must be defined. The precise estimation of fuel parameters is very demanding with respect to a vegetation complexity on a small research area and a diversity due to large elevation differences and protected character of the considerable portion of Slovak mountains. Therefore, an original methodology for the classification of forest vegetation in Slovakia and new fuel models were developed for selected localities [8,10,11]. Then reconstruction of the forest fire in the National Park Slovak Paradise in 2000 and analysis of the cause as to why people were trapped by the fire (6 people lost their lives) was carried out [8]. This paper presents our program module, which allows a dynamic 3D version of the forest fire simulation on real topography covered by "virtual forest". It is based on the output of FARSITE's fire spread simulation and it is developed for the purpose of 3D virtual reality of forest fire simulation. The described module is used for visualization of results of the reconstruction of the mentioned forest fire.

2.2 Models of simulations

The second group of modules in the Modelling tool is designed to generate Models of simulations. The following modules were created: CONVERTER, CLASSIFIER, COLOR SOTER, TIME SORTER and VIRTUAL SORTER and GENERATORS for static or for dynamic outputs. Our tool sorts these points according to the time and place, red faces for fire, blue faces for flood and brown faces for landslide. It starts from the point with the lowest time until the last one with the highest time. Input data for the second part of the Visualization tool are output data from simulators in different formats. Module CONVERTER converts them to singular format. Than data can input to CLASSIFIER module. CLASSIFIER sorts data and exports them to COLOR sorter and TIME sorter and some of them which are used for Virtual models to VIRTUAL sorter. Data from sorters input to GENERATORS. Generators are producing simulation models. Generator for static outputs produce **Static models.** Generator for dynamic outputs produce **Dynamic models.** Final visualization consists of terrain model and some simulation models. For example visualization of **fire spread in time** (see fig. 5) consists of TERRAIN.wrl and Fire spread.wrl files. **Flood in time** consists of TERRAIN.wrl and Flood.wrl files. Intensity of fire can be displayed statically.Outputs examples: see fig. 3.

Figure 3: Forest fire model of simulation.

2.3 Virtual reality models

The third group of modules is designed to generate models suitable for virtual reality applications. The following modules were created: DATA SORTER, VIRTUAL TERRAIN SORTER, VIRTUAL SIMULATION SORTER and several GENERATORS. Process of creation. Outputs from Virtual sorter together with Special data for Virtual models are exported to DATA SORTER. It sorts the data and exports them into VIRTUAL TERRAIN SORTER or VIRTUAL SIMULATION SORTER. Data from sorters are input to GENERATORS. They are producing a lot of virtual models like for example virtual terrain or forest, virtual buildings, virtual fire, or virtual flood. To create virtual models, the visualization tool needs *Special data for Virtual models* which can be variable. For example for generating virtual forest special data are *Grown maps*, or for generating some buildings they are *Project documentation* etc. The following example demonstrates using part of the Visualization tool for the creation of the Virtual forest fire. Firstly we created virtual forest from Grown maps which were provided to us by the forestry. For generation of virtual forest we used forest grown simulator Sibyla. This software was developed by the Forest faculty in Zvolen in Slovakia (See Fig. 4).

The Sibyla system is composed of ten modules. One of them is Forest GENERATOR Input data are values from Grown maps. There is included all required information for creating each quad of virtual forest. A huge database of textures included in Sybila enables us to create different looks of the forest (morning, day, evening, etc.). Then by using the Visualization tool the same terrain, which was firstly covered by ortophotomap is now covered by just prepared virtual forest (see Fig. 5). In the end red faces were replaced by virtual fire textures using Virtual fire generator module (see Figs. 6 and 7).

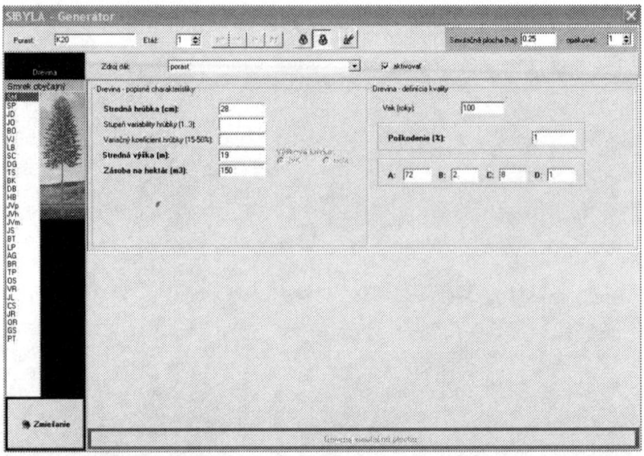

Figure 4: Sibyla forest generator.

Figure 5: Virtual forest.

3 Using the inVRs framework for collaborative visualization

In order to provide a deep insight of the displayed datasets the *inVRs* framework is used for immersive visualization [20]. *inVRs* offer the possibility to render virtual environments in stereoscopic 3D graphics on a variety multi-display

Figure 6: Virtual forest fire.

Figure 7: Forest fire during night.

output devices like the CAVE [5], curved screens or powerwall installations. The framework consists of independent modules for navigation, interaction and network communication, which are interconnected via its system core. Inside the core databases host objects and user information, like the datasets provided by pre-processing steps. To render the graphical display of *inVRs* applications OpenSG [21] is used as a scene graph. Users are able to navigate through the scenes depending on their chosen travel methodology and observe the scene as if they are standing in it. Different interaction methodologies can be chosen at the start up of the application. In the case of these visualizations switching of the different simulation steps is performed either automatically by starting an animation sequence or manually by pressing buttons to advance in the simulation. A valuable feature of the framework allows geographically dislocated users to collaboratively view and manipulate the datasets. To implement collaborative visualization the network module of the framework interconnects a set of clients in a peer-to-peer communication topology. The users are visualized by abstract avatars, which indicate the viewing orientation of the remote users.

Figure 8: Visualization in CAVE.

4 Visualization service

The Visualization Service is a common job submission service used to run the modeling by scripts. The service will produce visualization outputs and make them available for the portal. The results of 3D visualizations require an additional VRML browser plug-in to be installed. Visual service was used and evaluated during the MEDIGRID project. In the future it can be incorporated in related projects of the domain of disasters management [15,17].

5 Conclusion

Natural disasters simulation is a very complicated, challenging problem sensitive to the input data required. Therefore, intense research and development of sophisticated software systems and tools is extremely important for Natural disasters fighting management purposes. For example for Slovak forests, original methodology for forest vegetation classification and new fuel models have been developed and proper forest fire simulations related to the locality Krompla (National Park Slovak Paradise), where the large destructive fire appeared in 2000 and its reconstruction have been analyzed. These efforts induced the need of better auxiliary tools for 3D visualization of obtained simulation results and for animation of the forest fire spread. In this paper, new 3D visualization technique for real forest fire simulation and fire behavior and for flood and landslide modeling is described [18,19]. The importance is increasingly expanded for environmental problems [16].

Acknowledgements

This work was partially supported by Science and Technology Assistance Agency under the contract No. APVT-51-037902 and by the EU project MEDIGRID EU 6FP RTD GOCE-CT-2003-004044. And also is supported by RPEU project GRID_tools No. 0024-06 and VEGA project No. 2/7098/27.

References

[1] Juhas, F.: Forest Fire Slovak Paradise National Park (in Slovak), Report and Documentation of Department of Forest Fire Services, Spisska Nova Ves (2000).
[2] Xanthopoulos, G.: Forest fighting organization and approaches to the dispatch of forces in the European Union: results of the workshop survey, Proc. of the Int. Workshop on Improving Dispatching for Forest Fire Control (G. Xanthopoulos, ed.), Chania, Crete (2002), 143–153.
[3] Hille, M., Goldammer, J.G.: Dispatching and modelling of fires in Central European pine stands: New research and development approaches in Germany. Proc. of workshop.
[4] Finney, M.A.: FARSITE: Fire Area Simulator-Model, Development and Evaluation, Research paper RMRS-RP-4, USDA Forest Service (1998) Agronomic Institute of Chania, Crete, Greece (2001), 59–74.
[5] Viegas, D.X.: Surrounded by fire (in Portugal), Editorial Minerva, Coimbra (2004), 283 p.
[6] Viegas, D.X., Bibeiro, L.M., Silva, A.J. and Palheiro, P.: Analysis of S. Domingos accident, Proc. of the 4th Int. Conf. on Forest Fire Res. and Wildland Fire Safety, Luso, Portugal (2002) 18–23.
[7] Anderson, H.E.: Aids to determining fuel models for estimating fire behavior. USDA For. Serv. Gen. Tech. Rep. INT-122 (1982).
[8] Glasa, J., Halada, L.: Application of envelope theory for 2D fire front evolution. Int. Conf. on Forest Fire Research, Coimbra (2006).
[9] Richards, G.D.: An elliptical growth model of forest fire fronts and its numerical solution, Int. Journal for Num. Methods in Eng., 30 (1990), 1163–1179.
[10] Tuček, J., Schmidt, M., Celer, S.: Klasifikácia vegetačného krytu vo vysokohorských podmienkách z materialov DPZ s vysokým rozlíšením pri uplatnení apriorných poznatkov. Acta Facultatis Forestalis, vol. XLVII (2004), 89–102.
[11] Vida, T.: Metodika identifikácie a kvantifikácie palivových modelov pre simulovanie lesných požiarov, Diplom Work, Technical University in Zvolen, 2006.
[12] Viegas, D.X.: Surrounded by fire (in Portugal), Editorial Minerva, Coimbra (2004), 283 p.
[13] The EU project MEDIGRID EU 6FP RTD GOCE-CT-2003-004044.
[14] Hluchý, L., Habala, O., Nguyen, G., Šimo, B., Tran, V., Babík, M.: *Grid Computing and Knowledge Management in EU RTD Projects of IISAS.* In:

Proc. of 1st International Workshop on Grid Computing for Complex Problems – GCCP 2005.

[15] Šimo, B., Ciglan, M., Slížik, P., Mališka, M., Hluchý, L.: *Core services of heterogeneous distributed framework for multi-risk assessment of natural disasters.* International Conference on Computational Science – ICCS 2006, May 28–31, 2006.

[16] Halada, L., Weisenpacher, P., Glasa, J.: Reconstruction of the forest fire propagation case when people were entrapped by fire. Forest Ecology and Management, Vol. 234S, 2006, pp. 116, ISSN 0378-1127.

[17] Peter Slížik, Eva Pjorová, Martin Mališka, Ladislav Hluchý: Geovizualizations in Medigrid. International Workshop on Environmental Applications and Distributed Computing. – EADC October 16–17, 2006 Bratislava, Slovakia.

[18] Halada, L., Weisenpacher, P., Glasa, J.: Reconstruction of the forest fire propagation case when people were entrapped by fire. Int. Conf. On Forest Fire Research, Coimbra (2006).

[19] Ján Glasa, Eva Pajorová, Ladislav Halada, Peter Weisenpacher: Animation of Forest Fire Simulation. International Workshop on Environmental applications and Distributed Computing. EADC October 16–17, 2006 Bratislava.

[20] Christoph Anthes and Jens Volkert: "inVRs – A Framework for Building Interactive Networked Virtual Reality Systems". In International Conference (HPCC), pp. 894–904, Munich, Germany, September 2006.

[21] Dirk Reiners: "OpenSG: A Scene Graph System for Flexible and Efficient Realtime Rendering for Virtual and Augmented Reality Applications". PhD thesis, Technische Universität Darmstadt, Mai 2002.

Beach users' aesthetic and economic evaluation of a "minor change" to the hard engineering coastal defences at Wiseman's Bridge, Pembrokeshire, Wales

F. B. Blakemore, M. Burrell & S. D. R. Jones
Department of Science and Sport, University of Glamorgan, Wales, UK

Abstract

This study evaluated the visual impact of the new hard engineering coastal defences at Wiseman's Bridge, Pembrokeshire, Wales. This beach had existing hard coastal defences, road, car park and dwellings, which may have lead planners' to believe that the change would not impact upon beach users' enjoyment.

However, a change in coastal scenery was expressed by 79% of those surveyed, using the perceptual approach, whilst the beach classification was unaffected using two checklist models. The beach is primarily utilised by families for children's play and to enjoy the surroundings. Nearly half of the beach users had concerns with regard to the new coastal defences and were willing to pay (£1.62 per visit) to fund alternative defence methods for the remaining shoreline. A fifth of respondents were willing to pay (£1.91 per visit) for the new defences to be removed and replaced with a more natural method of sea defence. The preferred payment mode in both cases was a car parking fee. Thus, this study has found that even apparently minor changes in scenery impact markedly upon the recreational enjoyment and valuation of the beach experience.

Keywords: aesthetics, economic evaluation, willingness to pay, beach users, coastal defences.

1 Introduction

Due to current predictions of global sea-level rise and growing populations within the coastal zone, coastal defence is an issue that needs to be addressed at regional, national and global levels. Governments are being forced to come to terms with the reality of these issues, and subsequent environmental and socio-economic consequences. One aspect within a suite of UK coastal policy is Shoreline Management Plans (SMPs) these recommend five possible choices for protecting a coastline: hold the line: managed realignment; advance the line; limited intervention; and no intervention [1]. These SMPs also underline the importance of conserving and maintaining the natural environment, including the landscape [2]. The need for maintaining high standards in coastal landscape aesthetics has been outlined in a number of studies [3–6].

Coastal tourism is becoming the largest market sector of the tourism industry and accounts for 44% of all tourism in Wales [7]. The Welsh economy is therefore heavily dependant on coastal tourism [8]. Scenic quality and natural undeveloped beaches are increasingly being recognised as a major factor in attracting visitors to the Welsh coast. Blakemore and Williams [9] found scenery to be the second most popular reason given for visiting beaches in South-East Wales. Empathetic management of the coast is therefore needed in order to continue to attract visitors to coastal towns and villages.

1.1 Site location and description

Wiseman's Bridge is located on the South Pembrokeshire coast of West Wales, UK. Pebble sized clasts chiefly quartzites from the Coal Measure bedrock dominate the upper shore; the lower shore is composed of a sandy shingle mix, varying in composition and rock exposure. The beach has had several stages of coastal defence work over the years. This includes large boulders of Carboniferous Limestone and Dolerite forming a revetment type structure at the foot of the Wiseman's Bridge Inn, a small privately owned sea wall to the east of the Inn, and a large sea wall at the foot of the coastal path that links Wiseman's Bridge and Saundersfoot (*pers obs*). A car park and private dwellings are situated behind the coastal road that runs parallel to the beach, which has in recent years been overtopped during winter storms and pebbles have been deposited on the road [10]. In addition to this, coastal erosion had threatened to damage and undercut the road. The construction of a solid defence was completed in June 2004. The coastal defences have, in effect, replaced the majority of the berm crest at the top of the beach with an artificial seawall, leaving a limited amount of natural shoreline.

1.2 Aims of this study

This study utilised checklist and perceptual approaches to test the hypothesis that the minor changes in the visual impact of the new coastal defences at Wiseman's Bridge would not result in a change of classification in these categorisations. The way the beach is utilised by the public was determined and beach users' economic valuation elicited for the marginal gain in aesthetics, use and enjoyment of the beach with different coastal defences in place.

2 Beach evaluation

The Fuzzy Logic evaluation involved completion of a checklist consisting of 26 parameters (18 Physical; 8 Human) on the beach in question [11]. The value produced by this analysis is used to categorise beaches into five main classes [12]. Comparison of values before and after construction of the coastal defence at Wiseman's Bridge will highlight if any change in scenic quality has arisen according to this approach.

The BARE approach developed by Micallef and Williams [13], recognises five beach bathing area types. These are: Resort, Urban, Village, Rural and Remote. The five main issues of concern regarding beach user preferences and priorities are taken to be: Safety, Water Quality, Facilities, Scenery and Litter.

The perceptual survey took place over a seven-day period from the 19th to the 25th July 2004 at Wiseman's Bridge Beach. This period spanned the first complete week of the school holidays. Tourists were principally targeted because they are an important source of revenue in the area; as a result their opinions are likely to carry weight with coastal planners and developers seeking to promote Pembrokeshire as a tourist destination.

Respondents were chosen randomly, using a probabilistic systematic approach [14], i.e. every 3rd person or group was selected whilst walking along the beach. The nearest person in each group was asked to complete the questionnaire. People in the water were not approached due to the impracticality of hailing them from the shore. A total of 208 questionnaires were completed for full analysis. Sample sizes of 250-500 are usually recommended for open-format CV studies [14]; so the completed survey approached the lower margins of this suggested figure. Rejection rates were low (<2%).

2.1 Methods utilised in the perceptual survey

Respondents were asked to allocate a score (between –10 and +10) for scenic quality using photographs of Wiseman's Bridge, before and after the construction of the coastal defences (Figure 1). The use of photographs as environmental surrogates is widely practised in landscape scenic assessment [15], with many studies suggesting that judgments based on photographs are highly correlated with on site judgements of the same area using rating scales [16–19] or other response formats [20,21].

The photographs were taken at different times of the year ('before' in October 2003 and 'after' in June 2004) resulting a seasonal vegetation difference. Vegetation is considered a key component of landscape beauty [22] therefore these differences may bias results. The increased area of sand after construction, due to natural wave action, may also have influenced the responses. Studies on Italian beaches have found higher valuations for increasing beach area [23].

Two 'willingness to pay' (WTP) scenarios were posed: Respondents were first asked whether they would be willing to pay to fund an alternative method of coastal defence for the remaining one third of the beach, as yet unaffected by the defence works. The question included explanations of alternatives such as maintaining the natural shoreline or managed realignment.

Figure 1: 'Before' and 'after' images of Wiseman's Bridge which respondents were asked to score for scenic quality.

In the second scenario respondents were asked whether they would be willing to pay to fund the replacement of the new defences with an alternative scheme. Four options were described, supplemented by diagrams: revert to natural shoreline, removal of defences with beach replenishment, removal of defences with managed realignment and 'don't know/unsure'. No guidance on costs or on effectiveness of these alternatives was given as the aim was to consider aesthetic and beach use valuation only.

Both scenarios asked respondents which method of payment they considered most appropriate. Respondents were also asked to state how much they considered a reasonable contribution to pay per visit. No guide price was given as to what would be a reasonable charge for use of the beach in order to avoid influencing respondents and to eliminate starting point bias [24].

3 Results and discussion

The mean age of respondents was 48 years, with slightly more females being interviewed than males (57%). Females have been better represented than males in similar beach studies, such as that of Nelson and Botterill [25] (60% female respondents), who attributed this trend to mothers visiting the beach with their children.

Respondents were mainly tourists (87%), the largest contingent (56%) travelling from England. Visitors from South Wales formed the second highest proportion of beach users (24%). The small number of European visitors and a lack of tourists from further a field are consistent with the findings of Morgan[5], and reaffirm the notion that the South Wales coast is not a significant international tourist destination.

Day-visitors accounted for 13% of respondents. Most day-visitors, were encountered during the weekend, when they are less likely to be working. The majority of visitors were staying in the area for a week (49%) or a fortnight (23%).

3.1 Beach use at Wiseman's Bridge

The most popular priority stated for visiting the beach was 'For the children to play' (35%) indicates the importance of Wiseman's Bridge as a family beach. The second most popular priority (25%) for visiting Wiseman's Bridge was to 'enjoy the surroundings', indicating its aesthetic importance and that recreation can be an emotional as well as physical experience [26]. Walking was also rated highly (23%), with many respondents visiting Wiseman's Bridge to walk their dog, to stroll along the beach or to walk the coastal path. Less frequently given reasons for visiting the beach included swimming, sunbathing and water sports. Weather was a factor affecting these activities, as it was generally poor over the week of the survey. Swimming may be slightly under represented because visitors actively swimming were not approached.

3.2 Check list assessments

The majority of physical and human Fuzzy logic parameters were allocated high scores. The lowest scoring human parameters were access type (no buffer zone/light traffic) and utilities (two utilities present – sea wall and revetment). As sea walls and rock protection/revetments were previously present at Wiseman's Bridge, the building of the new defences did not alter the scoring of these parameters. These were the only parameters likely to change as a result of the building of the defences: The Fuzzy Logic assessment therefore indicates no change in scenic quality before and after construction of the sea defences. The assessment gave a value of 0.77, placing it in class 2 an attractive natural site with high landscape value.

Wiseman's Bridge does not meet the BARE specification to be a "village": as it is not associated with a small but permanent population reflecting access to organized but small-scale community services (such as a primary school(s), religious centre(s) and shop(s). Wiseman's Bridge best fits a "rural area" as it is located outside the urban/village environment. This classification was unaffected by the minor changes made to the beach defences. The changes to the beach defences only affects one of the five issues of concern regarding beach user preferences and priorities, that of Scenery. This is only marginally changed and so the BARE approach does not register the impact of the changes made at Wiseman's Bridge.

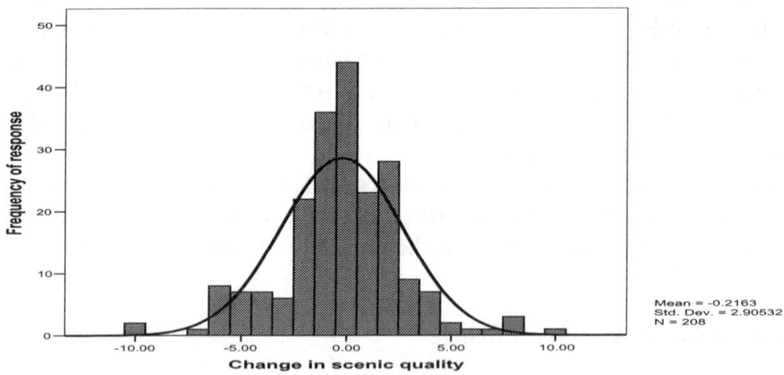

Figure 2: Variance in scenic quality ratings at Wiseman's Bridge as a result of construction of the new coastal defences.

3.3 Public perception of visual impact

Slightly more (43%) respondents gave Wiseman's Bridge a higher scenic quality score before construction of the new defences than (36%) after, however a considerable number believed scenic quality had not changed (21%). Respondents generally perceived little change in scenic quality, 73% changed by 2 points or less on a scale of minus 10 to plus 10, with a variation of zero (no change) being the most frequent.

This high central tendency coupled with a constrained range (Figure 2) deviates from a normal distribution (Kolmogorov-Smirnov test Z=2, p =0.001). A number of studies illustrate correlation between natural landscapes and higher aesthetic quality rating [4,18,27,28]. In the study the 'after' photograph was taken shortly following completion of the defences, with rock armouring at the base of the sea wall concealed by beach material, giving the defences a new and 'tidy' appearance (Figure 1). This picture also displayed a greater area of sand on the beach, which is likely to have improved the scores for the after photograph [23]. Results may indicate a higher preference for the pre-defence situation if the study is repeated using photographs taken a year or longer after construction, when the wall will have sustained inevitable wear and tear, and the rock armouring may be exposed.

Respondents were able to view the after scenario in three dimensions on-site as well as via a landscape surrogate (i.e. the photograph). Differences between depth of field of vision of photographs and human perception are considered by Bernaldez *et al.* [28] to be a serious limiting factor when comparing photo based and on-site visual assessments. This may possibly be a factor reducing respondents change in opinion between the two scenarios.

3.4 Contingent valuation; beach users 'willingness to pay'

The questionnaire revealed that 49% of respondents were willing to pay to use Wiseman's Bridge in order to fund an alternative method of coastal defence for the remainder of the beach, as yet unaffected by the sea wall. Tourists were more likely than day-visitors to express WTP to use the beach (tourists: 51% WTP; day-visitors: 35% WTP). This may be because local day-visitors use the beach more often than tourists and would have to pay frequently for a commodity that is presently free of charge. A charge for beach use may drive the local population to other areas. In contrast, tourists expect to incur costs for goods and services whilst on holiday. However, tourists are more likely to visit several places in one day so charges at each location may result in high cumulative costs. Therefore tourists visiting several areas may be less likely to visit Wiseman's Bridge if a charge is introduced for beach use. The mean value that the 49% of respondents who were WTP for beach use gave as a reasonable charge was £1.62. This is similar to the values found to maintain beaches in other published work such as Blakemore and Williams [29] (£1.03 per adult per visit) and King [30] (£1.78). No unreasonably high (protest) values were received (maximum value of £3.50 per day, one respondent).

In this study, chi-squared analysis indicated that there was no significant difference in WTP of respondents due to income level: Chi-squared = 4.67, $p = 0.458$ $df = 5$ and Cramer's $v = 0.153$. Comparison of each pair of income classes (classes as in figure 3) also indicated no significant differences in WTP.

The most popular method of payment for beach use was a car-parking fee (85%). The majority of beach users had arrived by car (66%). A number of respondents stated that they were visiting Wiseman's Bridge because; unlike nearby coastal resorts (Tenby and Saundersfoot), there is no car-parking fee. If this mode is used to raise revenue either: a daily parking ticket for any car park or charging an hourly rate for parking. would enable tourists to travel to several locations without paying excessive charges.

Respondents were also asked if they would be willing to pay to remove the present coastal defences and replace them with an alternative solution. This question received less support than the first scenario; only 21% of respondents expressed WTP. Comments made by respondents included: A general consensus amongst respondents was that the hundreds of thousands of pounds spent on the

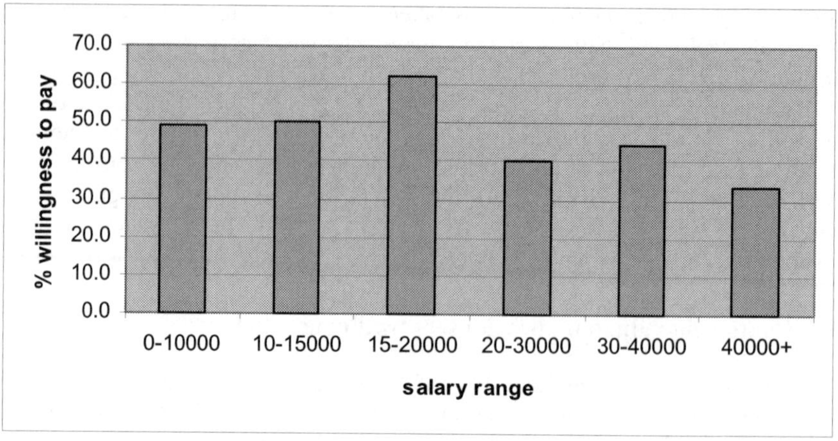

Figure 3: WTP for an alternative defence for remainder of the beach as a function of salary range.

structure should not be wasted and the best strategy now was to maintain the defences; many respondents did not believe the present defence was the best option, but were not concerned enough to pay to fund an alternative; others felt that Pembrokeshire County Council had taken this action because it was the only and or best option for defence of the road. The 21% of respondents WTP to remove the sea wall had a very strong opposition to the new defences. This is reflected in the higher amount they were WTP on average per trip (£1.91), compared to those WTP to fund an alternative solution for just the remainder of the beach (£1.62). The preferred method of payment was once again car-parking fees (76%). A similar proportion of both tourists and locals were WTP for this second scenario. The majority of respondents who expressed WTP to fund a replacement strategy considered either managed realignment (43%) or Beach replenishment (39%) to be the most suitable alternative option.

4 Conclusions

Wiseman's Bridge has been shown to be popular among families, with the most common reason for visiting the beach being 'for the children to play'. In conformity with other work, i.e. Morgan[5] and Blakemore and Williams [31,32], aesthetic quality has also proved to be important to beach users as 'enjoying the surroundings' was the second most popular reason for visiting Wiseman's Bridge. Coastal planners must be aware that recreation can be an emotion rather than an action performed and that the building of coastal defences can affect passive beach use of this kind.

This study suggests the new coastal defences at Wiseman's Bridge may have had a detrimental effect on the aesthetic quality of the coastal environment

The Fuzzy Logic approach developed by Ergin *et al.* [11,12] and the BARE system of Micallef and Williams [13] were unable to detect a change in scenic quality. This underlines the fact that affect on scenic quality was marginal. Thus, the concerns and negative responses obtained in this study may be magnified in a situation where hard engineering defences are constructed on a more natural beach.

The Contingent Valuation Method has demonstrated that around half of Wiseman's Bridge beach users are 'willing to pay' to preserve the remaining area of the upper beach from extension of the new defences. If a charging scheme was established for use of the beach, a car-parking fee is likely to cause least offence. The form of the parking fee introduced should not penalise tourists making various trips per day nor cause locals to use adjacent free beaches.

The second scenario, in which revenue generated would fund the replacement of the seawall with an alternative, highlighted that approximately a fifth of beach users felt a strong opposition to the defences, reflected in their higher valuation, £1.91 compared with £1.62 in the first scenario. Again this response is based on a marginal change to the pre-existing sea defences. Managed realignment, which is considered environmentally and economically viable under the Carmarthen Bay Shoreline Management Plan, was chosen by those respondents WTP as the most suitable alternative to the present defence. This demonstrates a degree of public competency in coastal defence issues and should encourage planners to allow the public to take a more active role in coastal defence decision-making.

Economic demand, beach use and public opinion regarding aesthetic impact are important when planning construction of coastal defences, especially in areas of scenic beauty, reliant on tourism such as Wiseman's Bridge

References

[1] DEFRA (2001) Shoreline Management Plans – A guild for coastal defence authorities. DEFRA Publications.

[2] ECUS (2003) *Guidance on coastal defence in relation to their landscape and visual impacts.* CCW Contract Science Report No. 531. Appleton, J. H. (1975) Landscape evaluation: the theoretical vacuum. *Trans. Inst. British Geography 66, p 120–123*

[3] Young, C., Barugh, A., Morgan, R. & Williams, A. T. (1996) Beach user perceptions and priorities at the Pembrokeshire Coast National Park, Wales, U.K. In: *Partnership in Coastal Zone Management,* Taussik, J. & Mitchell, J. (eds). *Samara Publishing Ltd., Cardigan.*

[4] Morgan, R. & Williams, A. T. (1999) Video panorama assessment of beach landscape aesthetics on the coast of Wales. *Journal of Coastal Conservation 5, p 13–22.*

[5] Morgan, R. (1999) Preferences and priorities of recreational beach users in Wales, U.K. *Journal of Coastal Research 15(3), p 653–667.*

[6] Somerville, S. E., Miller, K. L. & Mair, J. M. (2003) Assessment of the aesthetic quality of a selection of beaches in the Firth of Forth, Scotland. *Marine Pollution Bulletin 46(9), p 1184–1190.*

[7] WTB (2003) *Domestic (UK) tourism to Wales 2003* [online]: www.wtbonline.gov.uk

[8] Botterill, D., Owen, R. E., Emanuel, L., Foster, N., Gale, T., Nelson, C. & Selby, M. (2000) Perceptions from the periphery; the experience of Wales. In: *Aspects of Tourism in Peripheral Areas*, Brown, F. & Hall, D. (eds). Channel View Publications, Clevedon, p 7-38.
[9] Blakemore, F. B. & Williams, A. T. (1998) Public valuation of beaches in South East Wales, U.K. *Shore and Beach 4, p 18–23.*
[10] Royal Haskoning (2003) Technical Report for Pembrokeshire County Council (unpublished).
[11] Ergin, A., Karaesmen, E., Williams, A. T., Micallef, A., Karakaya, S. T. & Dedeoolu, M. R. (2003) Coastal scenery evaluation: applications of Fuzzy Logic mathematics at Turkish sites. *COPEDEC VI, Paper no. 164, Colombo, Sri Lanka*
[12] Ergin, A., Karaesme, E., Micallef, A. & Williams, A. T. (2002) A new methodology for evaluating coastal scenery: fuzzy logic systems. *AREA*.
[13] Micallef, A. and Williams, A.T. (2004) Application of a novel approach to beach classification in the Maltese Islands. *Ocean & Coastal Management*. 47, Issues 5-6, 2004, Pages 225–242. Morgan, R. (1998) A novel, user-based rating system for tourist beaches. *Tourism Management 20, p 393–410.*
[14] Bateman, I. J., Carson, R. T., Day, B., Hanemann, M., Hanley, N., Hett, T., Jones-Lee, M., Loomes, G., Mourato, S., Ozdemiroglu, E., Pearce, D. W., Sugden, R. & Swanson, J. (2002) *Economic Valuation with Stated Preference Techniques: A Manual*. Edward Elgar, Cheltenham.
[15] Hull, R. B. & Stewart, W. P. (1992) Validity of photo-based scenic beauty judgments. *Journal of Environmental Psychology 12, p 101–114.*
[16] Shuttleworth, S. (1980) The use of photographs as environmental presentation medium in landscape studies. *Journal of Environmental Management 11, p 61–76.*
[17] Nausser, J. I. (1982) Framing the landscape in photographic simulation. *Journal of Environmental Management 17, p 1–16.*
[18] Zube, E. H., Simcox, D. E. & Law, C. S. (1987) Perceptual landscape simulations: history and prospect. *Landscape Journal 6, p 62–80.1*
[19] Bosselmann, P. & Craik, H. K. (1989) Perceptual simulation of environments. In: *Methods in Environmental and Behavioural Research,* Bechtal, R., Marans, R. & Michelson, W. (eds). Van Nostrand Reinhold, New York. *p 162–190.*
[20] Shafer, E. L., Hamilton Jr., J. F. and Schmidt, E. A., (1969). Natural resource preferences: a predictive model. *Journal of Leisure Res. 1, p 1–19.*
[21] Palmer, J. F. (2004) Using spatial metrics to predict scenic perception in changing landscape: Dennis, Massachusetts. *Landscape & Urban Planning 69(2-3), p 201–218.*
[22] Appleton, J. H. (1975) Landscape evaluation: the theoretical vacuum. *Trans. Inst. British Geography 66, p 120–123.*
[23] Polomé, P., Marzetti, S. and Van der Veen, A. (2005) Economic and social demands for coastal protection, *Coastal Engineering*, 52 (10-11), pp. 819–40.

[24] Knetsch. J. L. (1993) *Environmental evaluations: some practical problems of wrong questions and misleading answers.* Resource Assessment Commission, Australian Government Service. Occasional paper No. 5, p 26.

[25] Nelson, C. & Botterill, D. (2002) Evaluating the contribution of beach quality awards to the local tourism industry in Wales – the Green Coast Award. *Ocean & Coastal Management 45(2-3), p 157–170.*

[26] Torkildsen, G. (1992) *Leisure and Recreation Management (3rd Edition).* Chapman & Hall, London

[27] Fines, K. D. (1968) Landscape evaluation: a research project in East Sussex. *Regional Studies 2, p 41–55.*

[28] Bernaldez, F. G., Ruiz, J. P., Benayas, J. & Abello, R. P. (1988) Real landscapes versus photographed landscapes: preference dimensions. *Landscape Research 13(1), p 10–11.*

[29] Blakemore, F. B. and Williams A. T. (2007) British tourists' valuation of Turkish beaches using Contingent Valuation and Travel Cost Methods. *JCR*, in press.

[30] King, O. H. (1995) Estimating the value of marine resources: a marine recreation case. *Ocean & Coastal Management 27(1-2), p 129–141.*

[31] Blakemore, F. B., Williams, A. T. & Ozhan, E. (2000) Tourist evaluation of Olu Deniz beach (Turkey) using Contingent Valuation and Travel Coast approach. *World Leisure 4, p 48–55.*

[32] Blakemore, F. B. & Williams, A. T. (2002). A comparison of tourist evaluation of beaches in Malta, Romania and Turkey. *World Leisure 2, p 29–41.*

The contribution of asset management to climate change policies

B. Bürgenmeier
Centre of Human Ecology and Environmental Sciences,
University of Geneva, Switzerland

Abstract

This paper deals with the three usual economic approaches to environmental protection, namely efficient markets, environmental economics and ecological economics. It focuses on a forth approach, which concerns the operational level of a business. This level is illustrated by an increasing number of criteria for global reporting, namely Social Accountability, ISO-norms etc, from the World Business Council for Sustainable Development. However, no commonly accepted standards exist. This paper shows what contributions "green" asset management makes to climate change policies.

Keywords: climate change, asset management, environmental economics, policy of environmental protection.

1 Introduction

Environmental protection in economics is simultaneously addressed by three different and somewhat controversial approaches.

The first approach analyses the change in relative prices over time in a pure market setting. Relative price changes (e.g. for energy of fossil origin) should express increased scarcity and induce technical progress and change in consumption patterns. If these changes cannot be observed, the market is not efficient. The consequent policy recommendation would be to strengthen competition and to reduce obstacles to the smooth functioning of markets.

The second approach brings the first one a step further as it is based on the concept of market failure. The environment is analysed as an externality to the market and can be understood either as a public good or as any other economic good which lacks a clear definition of property rights. The current policy

WIT Transactions on Ecology and the Environment, Vol 108, © 2008 WIT Press
www.witpress.com, ISSN 1743-3541 (on-line)
doi:10.2495/EEIA080131

recommendation concludes that the promotion of economic instruments is warranted. It can theoretically be shown that these instruments are protecting the environment at a lower cost than control and command instruments (Tietenberg [11]). This second approach is clearly in the field of environmental economics.

The third approach is developed in the field of ecological economics (Dore and Mount [2]). It introduces a change in perspective. The economy is external to the environment and has to be internalised into models of physical constraints to human activity, which cannot only be understood in terms of both economic and social dimensions. This paper argues that there is no model in sight that takes into account the social dimension in a satisfactory way. This dimension is probably the most misunderstood aspect of environmental issues. It raises fundamental questions about the aim and scope of economics. These questions are at the origin of the first reason why equity questions cannot be solved by models of growth (Constanza [1]).

Environmental issues have to be understood as a supplementary difficulty, added to the much older one of determining the "just price" which is tackled together with the "social question". The cumulative effects of the environmental and social consequences of human activities also extend the traditional approach of distributional justice to procedural and intergenerational justice with profound ethical implications (Hausman and McPherson [5]).

Increasing concern with social and environmental responsibility in business has led to the promotion of specially designed portfolios in asset management. This paper addresses this fact through five parts. The first introductory part recalls the ethical principles of the market in order to show that economic practices have been understood to be ethically sound, long before this trend actually occurred. Therefore, the second part shows that old problems have led to new questions and that the market economy stresses a never solved trade-off between efficiency and equity. The third part discusses the ethics of sustainable development and analyses this trade-off in relation to environmental protection. The fourth part is concerned with asset management, presents the main criteria and discusses the performance of sustainable portfolios. The fifth and final part concludes the reasons, explaining the trend towards social and environmental standards in asset management.

2 Welfare economics

That the functioning of markets is based on ethical principles is theoretically well grounded, but is sometimes forgotten in current business practices. These principles are well-known. Contracts have to be respected, competition has to be the binding rule of business, where transparency of all disposable information is granted and finally market access has to be unlimited for both businesses and consumers. Relative price changes lead to innovation through investment in research and development and to changes in consumer behaviour. The objective is profit maximisation on the level of the business and growth on the level of the nation.

Relative price changes have distributional consequences widely discussed in welfare theory throughout the history of economic thought. According to Amartya Sen, this "social question" stresses the fact that relative price changes exclude consumers with the lowest income and is intimately linked to the functioning of markets (Sen [8]).

Any attempt to analyse "economics as physics" is confronted with the normative aspects of the distribution of income and wealth. Standard economic approaches are bypassing the issue in focussing on output. The conclusion of welfare theory is to accept the normative content of distributional issues, and to show that it can be solved by economic growth (Feldmann [3]). In reducing sustainable development to sustainable growth in order to take into account equity issues, this strategy exclusively counts on modelling technical progress within the economic dimension.

Equity is only one aspect of the social dimension of sustainable development, which cannot be treated separately. It invites an understanding of complexity, instead of reducing it to models of simple causality. Equity is also interrelated with the economic and ecological dimension of sustainable development, while models of causality keep it separate. In focusing on ethics, the current discussion leaves out a long understood finding, namely that competition policy world wide is the best way of promoting ethical standards in business because it reduces all kinds of rents which are frequent in current economic practices. An international law of competition is lacking. The World Trade Organisation only reduces obstacles to free trade decided by States, but not of private businesses.

The fact that different aspects of equity have to be taken into account and several concepts of it co-exist makes it difficult for the business to clearly identify the extent of its responsibility. If only distributional equity is taken into account, it is also the most effectively addressed by economic growth. The specific characteristics of sustainable development are left out of the analysis.

One of the main findings of the 1987 report of the World Commission on Environment and Development [12] was the insight that there will no efficient policy for environmental protection if that policy did not also significantly reduce inequalities in income and wealth distribution in the world. This social dimension has not been studied systematically; most of the attention of the academic world has focused on the ecological and economic dimensions. Meanwhile, the business community has been confronted with an increasing separation of the real economy from finance, which is de-connected more and more from investment decisions for economic growth and not for financial performance.

3 Historical perspective

The origin of ethical investment goes back to the Great Depression in the thirties. Facing increasing unemployment, the trade-off between efficiency and equity appeared clearly. This historical reminder led to an increasing concern for public morality where the definition of ethical investment was determined by public health, arms and drugs. The recent trend for increased social responsibility of businesses only transforms this concern from a hidden agenda to a more transparently discussed issue. At the same time, public morality still condemned any treatment of risk and uncertainty during the nineteenth century as pure gambling. To bring such a treatment in the scope of economics was the task of insurance, which became a flourishing branch, and nowadays is more and more close to asset management and finance in general.

With the increasing conscience for environmental issues, ethical investment has finally been understood in relation to the concept of sustainable development including ethics in its social dimension. Table 1 lists the main aspects of this dimension and states the contextual framework of any modelling. The social dimension of sustainable development, like the concept of the market economy, is socially constructed. It is not so much the clear definition of this dimension which matters, but the image, the narrative content and the social symbols that are associated with it.

Table 1: Different aspects of the social dimension.

1	Social Institutions (Do institutions matter in human behaviour?) International and national institutional reforms and new bodies Legal framework (e.g. property rights) Monetary versus non-monetary evaluation
2	Life Style Analysis Consumption patterns (e.g. behavioural assumptions) Social images of Nature and technology (e.g. linguistic) Communication and Media
3	Social Cohesion Aggregation problems Government failures (intensity of preferences, Condorcet's paradox, lack of voting incentives)
4	Equity Normative content of the concept of equity Different forms of equity

3.1 Social institutions

The first group of arguments concerns the place of social institutions in economic analysis. Neoclassical theory both overemphasises and underestimates the importance of social institutions. It overemphasises them in assuming that social institutions are the rock foundation on which the economic calculus takes place. It underestimates them in assuming that individual preferences are given reflecting individual choices where social habits and institutions do not count.

In reforming the financial market in the perspective of sustainable development, not only are international institutions such as the Kyoto Protocol introduced, but also a new legal framework for emission trading and new evaluation methods for sustainable asset management.

3.2 Behavioural assumptions

Life Style Analysis can be approached in different ways. It is dominated by the theory of consumer choices based on cost-benefit analysis and the assumption of given preferences. Free rational choices are assumed in economic theory and there are as many behavioural assumptions as disciplines in the social and human sciences (e.g. Söderbaum [10]).

Different aspects of human behaviour are studied by different social sciences, but they can be divided into two main groups. The first one consists of consumer preferences when it comes to explaining why climate change policies have a more rhetorical than a practical support. The second one consists of social consumer habits but fails to clearly identify what is their origin. Human behaviour attributed to deep ecological concern can be expressed by lexicographical preferences, which are in contradiction to the assumption of transitivity underlying models of rational consumer behaviour (O'Connor and Spash [7]). There are attempts to clearly distinguish categories of behaviour of economic, ecological or social man, but these distinctions fail to answer the question of where the motivations for taking care of the environment come from.

3.3 Community

The third group of research in the field of the social dimension of sustainable development deals with the question of social cohesion. It offers a sociological perspective and a critique of economic approaches, which have to overcome an aggregate problem. If social welfare is just the sum of individual choices, where each knows best what is in his interest, it fails to grasp an important aspect, namely community building.

The fact that economic theory cannot link individual choice theory to social welfare has meant that some refer either to the representative individual which stands for all others, or to a social indifference curve without any link to individuals (Harsanyi [4]). This disconnection condemns any attempt to build models on microeconomic foundations in order to analyse social welfare functions and hence social equity issues. Those issues remain in the normative domain. Therefore, they are delegated to the political sciences in order to keep economic models consistent with the basic assumptions on which they are built.

4 Ethics of sustainable development

Ethics as a part of the social dimension of sustainable development combines the different findings of the above-mentioned groups. If natural resources are expressed as a source of welfare itself, the social image of Nature is used in national parks, created in the 19th century by the conservationist movement in the United States. Equity questions are then expressed by a more procedural form of justice. In this perspective, it is no longer possible to treat equity in the form of distributional justice only. Finally, if the social image of the environment is expressed in terms of garbage, society focuses environmental issues on recycling, and this refers to inter-generational equity.

The environment, seen as factor of production, as a source of well being, or as garbage, expresses different perspectives and addresses equity in different ways. Therefore equity cannot be a single expression for normative issues. It has to be understood in different terms making it even more difficult for a particular business to take equity into account.

The term "equity" stands for different interpretations of justice and can be expressed in different ways. It mainly covers distributional, procedural and intergenerational justice.

4.1 Distributional justice

Distributional justice intervenes on at least three distinctive levels:

- The first level deals with the distribution of income and wealth among individuals. It is closely linked to economic models based on a micro, business and household oriented foundation. The relevant question is to know which among several possible Pareto optimal solutions is the one which best satisfies distributional outcomes. This question can only be answered on normative grounds.
- The second level situates distributional justice among different countries. It is usually analysed within the theoretical framework of international trade, which is built on micro-foundations as well as reaffirming the usual finding that "more is better than less". Distributional issues are the consequence of environmental policies on an international scale such as the ones foreseen by the Kyoto Protocol. If they are addressed by separating their normative content from the positive analysis carried out as a formalised model, then they are exposed to the same shortcomings to which distributional issues on the individual level are exposed.
- The third level of distributional justice concerns inter-generational equity in terms of distribution issues, which introduce a new ethical yardstick in human activity. This distributional issue is unprecedented and is considered by many as the key feature of the concept of sustainable development. It refers to a fundamental choice. Either, this inter-generational distribution is ignored and conventional economic models continue to be extended to climate change; or it is not ignored and those models fail to grasp the very nature of the concept of sustainable development. The old "social question" is also asked in respect of future generations, which significantly increases the normative part of any economic reasoning.

4.2 Procedural justice

This second aspect of equity treats the way in which procedural justice shapes the economic process. The gearing principles of procedural justice are precautionary as well as intra-and inter-generational ethics. This procedural justice can be found in different stages of production, in transport and other intermediate activities, as well as in consumption. We recall four of them:

- At the first stage these principles intervene in the market of the factors of production. They are mainly used to justify energy and more generally all non-renewable natural resources saving programmes. They are also linked to the respect of social norms, which are set by ILO, but not always enforced (such as child work and minimum hygienic and health standards in resource extraction and pollution exposure).
- At the second stage, these principles design the social and ethical context in which a production function is set. Economic theory portrays such functions as a purely technological relationship between different factors of production. The technological part is expressed in models, the contextual one is not. We are confronted with two different kinds of approaches that cannot be put together in a unified and formalised framework.
- At the third stage which is defined by all intermediate activities between production and consumption, (such as merchants, financial intermediates, insurers and transporters), these principles set standards, which are often seen as obstacles to trade. The requirement to favour public transport systems instead of individual private vehicles is not only motivated by the internalisation of external effects, because it also has ideological implications. Public transport, for example, favours State intervention and relies more often on strategic planning procedures than on competitive markets.
- At the fourth stage, these principles are also shaping consumption. Herbert Simon has already insisted that rational consumer choices are bounded by asymmetric information (Simon [9]). We can now add that they are also bounded by precaution and ethics. Bounded rationality leads to cooperation and not so much to competition. This organisational perspective of consumer behaviour has lead to a theory of consumer behaviour, which analyses a household as a production unit. Though this theoretical extension of the classical consumer choice theory is based on preferences. This extension can be expressed by formal models and is also useful in order to take into account litter and waste; it does not fully include principles of procedural justice. Why should a consumer contribute to recycling his own waste if he pays taxes that are used for public garbage collection? The ethical incentive is the consumer's awareness to be part of a community, which is again an element that does not enter in formal models.

4.3 Intergenerational justice

Finally, this third aspect of equity is a particular form of procedural justice, which is extended to future generations. This extension introduces a radical change in perception and it is probably one of the major reasons why many scholars, mainly regrouped in ecological economics, are against the use of extended economic models applied to climate change without questioning the underlying methods of research. It is difficult to evaluate the long-term impact of research, which highlighted this radical change in perception (Martinez-Allier [6]). However, while new strands of thought are explored we can only promote pluralistic approaches coexisting and challenging formal modelling.

These approaches also insist on the cultural determinants of intergenerational justice and link this aspect of equity to the demographic factor intervening in environmental issues. If children are seen as descendants of a cult and moreover play an important part in perpetuating the related social order, intergenerational equity may have different meanings. It should to be understood in different cultural settings. Formal models are once again questioned as to their own contextual ingredients. Without generalising the cultural relativity of models, intergenerational, as a particular form of procedural justice, needs to be analysed with respect to cultural diversity. That does in no way mean a rejection of any attempt to understand climate change in building formal models. Rather it means seriously limiting their relevance for policy recommendations on a world wide scale.

5 Asset management facing ethics

The application of these different aspects of equity within the operational context of business is complex. One of the main reasons is given as the simple fact that macroeconomic growth objectives are not necessarily in the scope of businesses. Moreover, if models of economic growth are extended to environmental issues, equity problems occur because of the related change in relative prices designing winners and losers of the consequences of this change. The theory of welfare has extensively discussed the compensation criteria, no one can be considered as being scientifically neutral and value-free. The business has no clear guidance in order to know what society is considering as just. It can only deduce from the following attempts to translate ethical principles into operational concepts.

The first one is Agenda 21 that gives a concrete content to the ethical dimension of sustainable development. As a cost-benefit analysis for evaluating economic decisions and their consequences for the environment always leads to multi-criteria determination in order to properly carry out the required impact assessment, the weighting of the criteria is crucial for the outcome. This weighting should be the result of a participatory process of all stakeholders, who are concerned by an economic decision in a concrete and therefore mostly local context. Businesses cannot ignore this evolution and are bound to collaborate more and more with local groups.

The second is the strict application of the BAT principle, namely to produce and to distribute goods and services according to the best available technique in respect to the environment. Those businesses that are technologically advanced from an environmental point of view have a strategic advantage over their competitors reflected in the appreciation of the stock market.

The third way to translate ethical principles into concrete action follows logically from the two preceding ones. Participation and stock market evaluation is at the root of green asset management and of numerous evaluation procedures in order to determine how the business handles its relationship with the stakeholders from the civil society and how to determine green investment. Table 2 lists some examples of such an endeavour.

These different attempts to evaluate business practices in order to promote the social and ecological responsibility of the business are the precondition for creating portfolios for green and social asset management.

However, the success of that kind of assessment management may be more mundane. The performance of such portfolios is as good as and sometimes better than ordinary portfolios. Table 3 compares the performance of a "green portfolio" composed of only shares of the members of the World Business Council of Sustainable Development to the usual benchmarks in the US and in Europe at the end of August 2005.

Table 2: Examples of norms and standards of evaluation.

Organisation	Main objectives
Global reporting	Global Reporting Initiative, Sustainability Reporting Guidelines on Economic, Environmental and Social Performances
Accountability	Promotion of standard AA 1000 for Governance (applied by Novo Nordisk, The Body Shop, British Telecom e.g.) Institute of Social and Ethical Accountability
ISO	Promotion of norm ISO 1431 by the International Organisation for Standardisation Council on Economic Priority Accreditation Agency
CEPAA	Promotion of a Social Accountability standard ISP 8000
WBCSD	World Business Council for Sustainable Development Guide for businesses

Table 3: Performance.

Market	1 Year	5 Years
Standard & Poor's 500	12.5%	-16.4%
"green" portfolio	10.8%	14.1%
Dow Jones Europe		
Stoxx 50	1.4%	-29.1%
"green" portfolio	5.1%	17.9%

Source: Dexia and BCSD.

These figures illustrate that, with the exception of a one year performance in the US (Standard & Poor's 500), the comparisons are in favor of the "green" portfolios.

6 Conclusion

The tendency towards corporate governance, social norms and indicators for sustainable development will also increase in the future for two reasons. The first is internal, the second external to the business.

The internal reason is linked to the motivational behaviour of employees. In order to increase the motivation to work, the business has to send out a clear message to its own employees, that it cares for the environment and that it is concerned by social issues.

The external reason is linked to the behaviour of consumers. The legitimacy of a business is crucial for a convincing marketing. The increasing awareness for environmental and social concern is therefore shaping business behaviour and its evaluation by the stock market.

Are new business ethics necessary or is it enough to strengthen the well-known ethical principles of the market? It is difficult to answer this question and to guess future trends, but the traditional economic rationality will be completed more and more by the precautionary principle, competition by local cooperation, mobility by social rooting and finally the implicit social contract within generations will be extended to intergenerational responsibility. Modern economic teaching should fully integrate these aspects right from the beginning.

References

[1] Constanza, R., Ecological Economics: The Science and Management of Sustainability, Columbia University Press: New York, 1991.
[2] Dore, M.H.I. & Mount, T.D. (eds). Global Environmental Economics, Equity and the Limits to Markets, Blackwell: Oxford, 1999.
[3] Feldmann, A.M., Welfare Economics and Social Choice Theory, Kluwer: Boston, 1983.
[4] Harsanyi, J. (1955), Cardinal welfare, Individualistic ethics, and interpersonal comparisons of utility. Journal of Political Economy, No 63, 1955.
[5] Hausman, D.M. & McPerson, M.S., Taking ethics seriously: Economics and contemporary moral philosophy. Journal of Political Economy, No 63, 1993.
[6] Martinez-Allier, J., Ecological Economics, Blackwell: Oxford, 1987.
[7] O'Connor, M. & Spash, C.L., Valuation and the Environment, Theory, Method and Practice, Edward Elgar: Aldershot and Brookfield, 1999.
[8] Sen, A., On Economic Inequality, Oxford University Press: Oxford, 1973, expanded edition with a substantial annexe by J. E. Foster and A. Sen, 1997.
[9] Simon, H.A., Models of Bounded Rationality, The MIT Press: Cambridge and Massachusetts, 1982.
[10] Söderbaum, P., Ecological Economics, Earthcan: London, 2000.
[11] Tietenberg, T.H., Innovation in Environmental Policy, Economic and Legal Aspects of Recent Developments in Environmental Enforcement and Liability, Edward Elgar: Aldershot and Brookfield, 1992.
[12] World Commission on Environment and Development, Our Common Future, Oxford University Press: Oxford and New York, 1987.

Market-based instruments in South Africa: a review

A. Nahman, L. Godfrey & R. Wise
Council for Scientific and Industrial Research, South Africa

Abstract

Sustainable development presents particular challenges for developing countries, where the need for rapid economic development to overcome poverty and inequality often makes environmental protection a low priority. Market-based instruments (MBIs), which make it possible for environmental considerations to be built into everyday economic decision-making, and for environmental objectives to be achieved at the lowest possible cost to society, may be appropriate policy instruments for addressing these types of challenges. However, despite their potential advantages in the developing country context, MBIs may not always be appropriate for developing countries, and should not be seen as panaceas that can simply be transferred to developing countries without considering local circumstances. By way of example, this paper reviews the use of MBIs for solid waste management in South Africa, and presents results from a survey of waste management authorities regarding the opportunities and constraints associated with their implementation. It is found that a number of fundamentals need to be put in place before MBIs can be implemented. However, this does not imply that the potential of MBIs should be ignored. Indeed, it is proposed that MBIs can be designed and implemented in an innovative, incremental way whereby capacity and experience are gradually developed over time.

Keywords: market-based instruments, solid waste management, developing countries, sustainable development, panaceas.

1 Introduction

The challenges associated with sustainable development are particularly difficult and complex in developing countries, where trade-offs between economic, social

WIT Transactions on Ecology and the Environment, Vol 108, © 2008 WIT Press
www.witpress.com, ISSN 1743-3541 (on-line)
doi:10.2495/EEIA080141

and environmental objectives are unavoidable. Given the need for rapid social and economic development to overcome poverty and under-development in most developing countries, decision makers tend to prioritise social and economic development agendas, often at the expense of environmental integrity. A more balanced approach is therefore required, whereby inter- and trans-disciplinary tools, methods and frameworks are used to facilitate the decision-making process (Burns *et al* [1]). This is particularly important where social, economic and environmental variables are seemingly incommensurate.

Environmental economics is an inter-disciplinary science devoted to the resolution of these types of trade-offs. It is based on the idea that all forms of capital, including social and natural capital, can be valued in economic terms, facilitating the resolution of tradeoffs between economic, social and environmental objectives. In terms of policy, environmental economists advocate market-based approaches to resolving these trade-offs, whereby environmental objectives can be achieved at minimum cost to society, thus easing the economic burden of environmental protection. It provides insight into the impact of incentives on people's decision making; and can then construct incentives to ensure that people's behaviour leads to sustainable outcomes. Environmental economics, and market-based instruments (MBIs) in particular, therefore seem well-suited to addressing the challenges associated with sustainable development. It would thus seem that an environmental economics perspective is particularly important in developing countries; where limited resources, and the need for economic growth and development, imply that cost-effectiveness should be a key criterion for environmental policy.

However, economists have tended to see MBIs as a panacea that can be applied to all environmental problems in all contexts. There has therefore been a tendency for economists to advocate the use of sophisticated, developed country instruments in developing countries, which may not necessarily be ready for them, or which at least require that the context be taken into account (Bell [2], Bell and Russell [3], O'Connor [4], Russell and Vaughan [5], Ostrom *et al* [6], Brock and Carpenter [7]). An underlying characteristic of this 'panacea thinking' is that "the set of preferences, the possible roles of information, and individual perceptions and reactions are assumed to be the same as those found in developed Western market economies" (Ostrom *et al* [6]). This is often not the case, resulting in misguided policies that often do more harm than good (Ostrom *et al* [6], Brock and Carpenter [7]).

The aims of this paper are (1) to assess the opportunities for, and constraints to, the implementation of MBIs in the solid waste sector in South Africa (SA); and (2) to suggest an approach to the implementation of MBIs that takes the developing country context into account. Section 2 introduces the concept of MBIs and reviews MBIs that have been used in the solid waste sector internationally. Section 3 presents the South African context, while Section 4 briefly describes the survey methodology used to assess opportunities for and constraints to the implementation of MBIs in SA. Section 5 presents the results, while Section 6 suggests a possible approach for implementing MBIs in developing countries.

2 Market-based instruments in theory and in practice

Environmental problems have traditionally been addressed using command and control (CAC) regulations, which regulate behaviour directly by prescribing specific legislation and standards which must be achieved, and enforcing compliance through the use of penalties and fines (Perman *et al* [8]). In developing countries, including SA, CAC mechanisms still dominate environmental policy (Pearce and Turner [9], National Treasury [10]).

By contrast, MBIs, such as environmental taxes and subsidies, seek to change behaviour *indirectly*; by changing relative prices (and hence incentives) that individuals and businesses face. These instruments have grown in importance in developed countries since the 1980s; and there has also been growing interest in some developing countries in the use of such instruments (Bell and Russell [3], Pearce and Turner [9], Inter-American Development Bank [11]).

Various types of MBIs can be used in the field of solid waste management (SWM). These include product and input taxes, deposit-refund schemes, and quantity-based user charges.

Product and input taxes aim to reduce waste generation by increasing the price of environmentally significant products and inputs; thereby decreasing demand or making alternative products or inputs more attractive. Internationally, they have been applied to tyres, motor vehicles, batteries (particularly car batteries), non-recyclable containers (plastic, glass, metal and paper; particularly non-returnable beverage containers), non-biodegradable plastic bags, lubricant oils, and fuels (Forum for Economics and the Environment [12], United Nations Environment Program [13]). They are generally applied at a national level.

Deposit-refund schemes essentially combine a product tax (the 'deposit') and a subsidy ('refund'). The deposit is paid upon purchase and is refunded upon return of the used product or packaging for recycling or re-use, thereby providing an incentive to return recyclable or reusable items rather than throw them away (Forum for Economics and the Environment [12], Pearce and Turner [14]). They are most commonly used for beverage containers, although they have also been used for tyres, batteries, and even cars (e.g. in Sweden and Norway) (Inter-American Development Bank [11]). They are generally applied at a local level.

Finally, quantity-based user charges for collection or disposal aim to provide an incentive to reduce waste generation or disposal at the margin, or to divert waste from landfill to recycling or re-use. For example, charges for municipal waste collection services can be based on the volume or weight of waste collected. Quantity-based collection charges have been used by some municipalities in Switzerland (and some other European countries), South Korea, the United States and Canada (Inter-American Development Bank [11], Pearce and Turner [14], Reschovsky and Stone [15], Choe and Fraser [16]). Variable landfill fees (based on weight) are also common, although these generally fail to affect the incentives of waste generators themselves and therefore have little influence on the quantity of waste generated. Quantity-based charging is usually applied at the municipal level.

MBIs appear to have some important advantages over CAC in the developing country context, including cost effectiveness; promotion of economic efficiency; incentives for innovation; the potential for self-regulation; and the potential for revenue generation, which can be used to finance environmental expenditures or to reduce distortionary taxes elsewhere in the economy, such as on labour, thereby generating a 'double dividend' (National Treasury [10]).

Nevertheless, MBIs require the fulfilment of a number of pre-conditions. Firstly, they require "reasonably well functioning markets with adequately defined property rights, the presence of private enterprise motivated to reduce costs, some degree of competition, competent judicial systems, and limited price distortions" (Inter-American Development Bank [11]). Developing countries are generally characterised by poorly functioning markets and/or market failure, whereby these conditions fail to hold.

Secondly, MBIs can be administratively demanding on government at all levels, and require sufficient institutional capacity in terms of acquiring relevant management information, monitoring compliance and illegal activities, and enforcement (United Nations Environment Program [13]). Many developing countries lack the necessary institutional capacity (Bell and Russell [3], Russell and Vaughan [5]). Thus, institutional capacity building is an important precursor to the implementation of MBIs (Pearce and Turner [14]).

Finally, there must be political will to address environmental problems as a priority issue, particularly given the likelihood that many MBIs (particularly taxes and charges, which may impact negatively on competitiveness and equity) will be politically unpopular. However, there are various ways of alleviating the negative impacts associated with taxes so as to ensure that the overall impact of the tax is neutral; such as tax shifting, revenue recycling and transfer payments.

3 The South African context

There are few MBIs in SA with explicit environmental objectives. Although taxes and charges exist in the transport, energy, water and waste management sectors, these are largely intended for cost-recovery or revenue-raising purposes, with any environmental outcomes being achieved only incidentally (National Treasury [10]). Currently, the only product tax in SA with explicit environmental objectives is the plastic bag levy, although there is also a proposed levy on tyres and potential for the expansion of product taxes to such items as packaging, batteries, and electronic equipment (National Treasury [10]). There are also a number of locally operated deposit-refund schemes in SA, relating largely to glass and plastic beverage containers. Again, these can potentially be expanded to include other products (National Treasury [10]).

Charging for waste collection in SA varies widely between municipalities, with some municipalities funding waste collection services entirely through property taxes, and others almost entirely through user charges, and still others using some combination thereof. In the case of user charges, some are based on plot size or property value, while others are based on the frequency of collection or number of containers collected (National Treasury [10]).

4 Methodology

Face-to-face and telephonic semi-structured (questionnaire-based) interviews were conducted with 18 individuals from 15 municipalities and private waste management companies across SA between October and December 2007. Discussions were also held with representatives of the National Treasury, although this paper confines itself to the results of the discussions with local authorities. Copies of a summary report on MBIs for SWM, as well as the survey questionnaire, were distributed prior to the interviews. Interviewees were encouraged to read the report and begin responding to the questionnaire prior to the interview. During the interview itself, interviewees were presented with a brief summary of MBIs and the issues involved, after which semi-structured discussions were held based on the questionnaire. Interviewees were also encouraged to raise other issues not dealt with in the questionnaire. Interviewees were then asked to complete the questionnaire in their own time if necessary, and to provide information regarding their waste collection tariffs and landfill tipping fees (where applicable).

5 Results

Respondents were first asked questions regarding their current waste management charging practices. Most (80%) of the municipalities and companies in the sample claimed to have some form of variable or quantity-based tariff structure, with two municipalities using fixed rates and one not charging for waste collection at all. However, of those claiming to use variable or quantity-based charges, only in one case are charges based on the actual weight or volume of waste collected. In all other cases, charges are based on the number of containers collected. The problem is that, in most cases, the unit by which waste generators are charged is one 240 litre 'wheelie' bin, or between two and four 85 litre bins or bags, or part thereof. These large units fail to encourage waste minimisation, as waste generators are charged the same amount irrespective of how full the bins are, or of how many bags (within the prescribed limit) are collected; such that there is still no incentive to reduce waste generation at the margin (Fullerton and Walls [17]). Only in two cases are charges based on a unit of one 85L bin, which is clearly preferable to charges based on larger units; although, again, charges are the same irrespective of how full the bin is. There have been some proposals to move toward a weight-based charging system, although such a system requires fairly sophisticated equipment and monitoring capabilities. Furthermore, quantity-based user charges are likely to lead to an increase in illegal dumping as a means of avoiding the fees.

Sixty-six percent of municipalities and companies in the sample have landfill tipping fees based either on weight (where a weighbridge is available) or volume. Again, however, the units involved tend to be large (such as per tonne of waste, or based on vehicle size). Furthermore, marginal disposal costs are not generally passed on to waste generators in the form of marginal waste collection costs, such that, again, little incentive for waste minimisation is created.

Furthermore, the true costs of landfilling (including externalities, e.g. health, social and environmental costs) are not currently built into landfill tipping fees in SA. Indeed, none of the municipalities interviewed claimed to cover external costs in their waste collection or disposal fees, although one claimed to be looking into this issue. The result is an artificially low cost of landfilling, which makes recycling an unattractive alternative. Thus, correct pricing of landfilling (increasing landfill fees to reflect the full costs associated with this activity) is an important priority. However, as is the case with quantity-based collection fees, increasing landfill tipping fees is likely to lead to an increase in illegal dumping. Thus, as mentioned by several respondents, changing social attitudes regarding appropriate waste management behaviour, through education and awareness programmes, is an important precursor to the implementation of MBIs.

Respondents were then asked questions regarding the opportunities and constraints associated with implementing MBIs for SWM in SA. Seventeen out of the 18 individuals (94%) felt that MBIs should be implemented in the field of SWM in SA, with only one respondent unsure. However, opinion varied as to the time frame over which they should be implemented. Eleven respondents (61%) felt that at least some MBIs, or some aspects of MBIs, could be implemented over the short term (0-5 years), with only 6 respondents (33%) arguing that MBIs should only be implemented in the medium term (5-10 years); and only one respondent arguing that they should only be implemented in the long term (over 10 years). Twenty two percent of respondents argued that different MBIs, or different aspects of MBIs, should be implemented over different time frames.

Respondents saw a variety of opportunities in implementing MBIs. All respondents thought that MBIs could be effective in achieving some reduction in waste generation or in diverting waste from landfill to recycling. However, only six respondents (33%) thought that MBIs should be implemented with the sole purpose of achieving these outcomes. Two saw income generation as an important secondary benefit; while 8 respondents (44%) saw the two outcomes (changing behaviour and generating income) as equally important. Only one respondent thought that income generation should be the main purpose of MBIs.

Other opportunities mentioned by respondents included the potential of MBIs to reduce the need for landfill space and prolong the lifespan of landfill sites; their potential to stabilise prices of recyclables and thus stimulate and stabilise viable and sustainable markets for recyclables; the socio-economic benefits associated with recycling, such as local economic development and the creation of job opportunities in the recycling market; improved environmental awareness; skills and technology development; the potential to encourage private investment; and the potential efficiency gains associated with municipal waste management departments being run more like businesses.

However, a large number of obstacles to the implementation of MBIs for SWM in SA were mentioned by respondents (Table 1). Only seven respondents (39%) felt that MBIs could be easily integrated within the current SA policy framework, while three were unsure. Eight respondents (44%) felt that MBIs may not be easily integrated into the current policy framework, although half of these mentioned that this would change when the new Waste Management Bill is enacted. According to these latter respondents, MBIs could be added as regulations within the framework provided by the Bill.

Table 1: Obstacles to implementation of MBIs for SWM in SA.

Question	Yes	No	Unsure	Other
Can MBIs be easily integrated into the SA policy framework?	39%	22%	17%	When Waste Bill is enacted 22%
Is SWM seen as being of sufficient priority for implementation of MBIs?	22%	61%	0%	Not yet but starting to become so 17%
Does government have the necessary enforcement capacity for MBIs?	6%	83%	6%	When Waste Bill is enacted 6%
Does your organisation or municipality have the required capacity for MBIs?	22%	50%	11%	Not yet but can be developed 17%

Furthermore, 11 respondents (61%) believe that waste management is not seen as enough of a priority, either at local or national government level, to enable implementation of MBIs. Four respondents specifically mentioned lack of political will as an important obstacle to their implementation. Likewise, seven respondents (39%) cited lack of waste education and awareness among the public as an obstacle. Thirteen respondents (72%) mentioned raising awareness and changing mindsets, e.g. through environmental and waste education or advertising campaigns, as a fundamental precursor to the implementation of MBIs, or as being a necessary complement to the implementation of MBIs, or even as being a key instrument for changing behaviour in its own right.

Another important obstacle mentioned by nearly all respondents was the lack of administrative and enforcement capacity at all levels of government. Only one respondent felt that national government had the necessary enforcement capacity to implement MBIs at the national level in SA, with another respondent unsure. The vast majority (89%), however, doubted the existence of sufficient capacity at national level. Of these, one respondent argued that an enabling environment for such enforcement would be created once the Waste Management Bill is enacted.

Similarly, 12 respondents (67%) felt that there was a lack of capacity within their municipalities or organisations for the implementation of MBIs at the local level (responses to this question did not vary significantly between municipalities and private companies). Of these, three (all municipalities) claimed that such capacity could be developed, although this would require an increase in funding and/or staffing. Four respondents claimed that such capacity already existed in their municipalities or organisations, while two were unsure.

Respondents also mentioned the need to create an enabling environment (e.g. provision of bags and drop-off centres, or even kerbside pickup, for recyclables); the difficulty of monitoring waste generators due to the lack of waste licensing and data (e.g. due to lack of a waste information system); low prices of recyclables, which makes use of virgin materials more attractive; the lack of uniformity and stability in the prices of recyclables (and hence in the market for recyclables); the costs associated with transport, monitoring and enforcement;

lack of access to basic waste services; the negative impacts of further taxes on the poor and on business, and thus the likelihood of resistance from communities and business; other difficulties associated with the enforcement of new policies; lack of staff (particularly skilled staff), funding, vehicles, and other necessary resources; the budgeting process, which makes it difficult for municipal waste management departments to re-invest income into improved SWM; the complicated nature of these instruments, especially in light of the current inability of many municipalities even to deal with basic billing and cost recovery issues (for example, only 43% of municipalities in the sample who charge for waste collection are able to cover the full operational costs of waste collection services through these charges); and lack of the necessary enabling legislation and legal framework (at least until the Waste Management Bill is enacted).

A number of potential problems were mentioned with regard to specific instruments. In the case of taxes and charges, concerns were raised regarding the impact of taxes or charges on (and thus resistance from) poor communities and businesses; and as to how revenues will be used (e.g. the case of the plastic bag levy, revenues from which are not properly channelled toward the stated purpose of stimulating recycling). In the case of quantity-based charges specifically, concerns were raised regarding possible incentives for illegal dumping created by such charges; the additional data requirements, complexities and costs associated with monitoring and billing (adequate waste data and monitoring/billing capacity are already severely lacking); and regarding the need to change the mindset that waste services should be free.

6 Summary and conclusions

In general, almost all respondents felt that MBIs could lead to reduced waste generation and increased diversion of waste from landfill to recycling, and that they should eventually be implemented in the field of SWM in SA, although opinions varied as to the appropriate timeframe for implementation. The general consensus was that a number of fundamentals had to be in place first, including:

- enactment of the Waste Management Bill, which will create an enabling environment for enforcement and provide a framework within which MBIs can be implemented;
- political will (government must see waste management as a priority);
- education and awareness (waste management must be seen as a priority among business and communities; to encourage waste minimisation and recycling, reduce illegal dumping, and promote acceptance of MBIs);
- development of capacity at all levels of government (for administration, monitoring and enforcement of instruments, and billing for services);
- increased access to resources for waste management departments (to develop capacity, recover costs, and improve SWM services);
- correct pricing of landfilling, such that external costs are accounted for;
- waste licensing and data (e.g. through a waste information system); and
- infrastructure for extension of basic waste services, improvement in existing services, and to enhance the convenience of recycling.

Thus, it may not be appropriate to push for sophisticated MBIs before these fundamentals are in place. However, an existing lack of institutional capacity should not be taken to imply that there is no place for MBIs in developing countries. Instead, institutional limitations must be taken into account in the selection, design and implementation of such instruments.

For example, some instruments may be simpler to implement and administer than others. In this study, respondents indicated concern regarding the lack of monitoring and enforcement capacity at municipal level, especially for the billing of waste services and the monitoring of illegal dumping in the case of quantity-based user charges. This implies that simpler instruments that are more easily enforced and that don't create perverse incentives, such as product taxes and deposit-refund schemes, should be preferred over quantity-based charges.

Furthermore, an existing lack of institutional capacity can be taken into account in the design and implementation of instruments. Policy can be designed in such a way that instruments are implemented incrementally, beginning with relatively simply instruments and becoming increasingly sophisticated as institutional capacity grows (Bell and Russell [3], Russell and Vaughan [5], Pearce and Turner [9]). It is also important to develop a culture of compliance whereby compliance becomes the norm and illegal dumping becomes socially unacceptable (Russell and Vaughan [5]). For example, to start with, it is likely to be easier to place a tax on products at the point of sale, or to monitor waste entering landfill sites or generated by large producers; rather than attempting to monitor the quantity of waste generated by individual households, as well as illegal dumping. It may also be possible to implement deposit-refund schemes, or expand existing schemes to cover other types of products (Bell and Russell [3]). However, as monitoring capacity and a culture of compliance develops, it will eventually become easier to monitor household waste generation, while (ideally) avoiding having to deal with illegal dumping. Further research is therefore required regarding the design of instruments that take developing country circumstances into account and that can develop over time as capacity grows.

References

[1] Burns, M., Audouin, M. and Weaver, A., Advancing sustainability science in South Africa. *South African Journal of Science*, **102**, pp. 1–6, 2006.

[2] Bell, R.G., *Choosing environmental policy instruments in the real world.* CCNM/GF/SD/ENV(2003)10/FINAL. Paris. Organisation for economic cooperation and development, 2003.

[3] Bell, R.G. and Russell, C., Environmental policy for developing countries. *Issues in Science and Technology*, **18(3)**, pp. 63–70, 2002.

[4] O'Connor, D., Applying economic instruments in developing countries: From theory to implementation. *Environment and Development Economics*, **4**, pp. 91–110, 1998.

[5] Russell, C.S. and Vaughan, W.J., The Choice of Pollution Control Policy Instruments in Developing Countries: Arguments, Evidence and

Suggestions. *International Yearbook of Environmental and Resource Economics*, VII ed., Edward Elgar: Cheltenham, U.K., 2003.

[6] Ostrom, E., Janssen, M.A. and Anderies, J.M., Going beyond panaceas. *Proceedings of the National Academy of Sciences, USA*, **104(39)**, pp. 15176–15178, 2007.

[7] Brock, W.A. and Carpenter, S.R., Panaceas and diversification of environmental policy. *Proceedings of the National Academy of Sciences*, **104**, pp. 15206–15211, 2007.

[8] Perman, R., Ma, Y., McGilvray, J. and Common, M., *Natural resource and environmental economics*, Pearson Education: Harlow, 2003.

[9] Pearce, D. and Turner, R.K., *Economics and solid waste management in the developing world.* CSERGE working paper WM 1994–05. Centre for Social and Economic Research on the Global Environment, 1994.

[10] National Treasury, *A framework for considering market-based instruments to support environmental fiscal reform in South Africa*. Pretoria. National Treasury, 2006.

[11] Inter-American Development Bank, *Economic instruments for solid waste management: global review and applications for Latin America and the Caribbean*. Washington, D.C. Inter-American Development Bank, 2003.

[12] Forum for Economics and the Environment, *Training manual for the Forum for Economics and the Environment.* http://www.econ4env.co.za/training/training.html 2002.

[13] United Nations Environment Program, *Solid waste management.* http://www.unep.or.jp/ietc/Publications/spc/Solid_Waste_Management/index.asp. 2005.

[14] Pearce, D. and Turner, R.K., Market-based approaches to solid waste management. *Resources, Conservation and Recycling*, **8**, pp. 63–90, 1993.

[15] Reschovsky, J.D. and Stone, S.E., Market incentives to encourage household waste recycling: Paying for what you throw away. *Journal of Policy Analysis and Management*, **13(1)**, pp. 120–139, 1994.

[16] Choe, C. and Fraser, I., The economics of household waste management: A review. *The Australian Journal of Agricultural and Resource Economics*, **42(3)**, pp. 269–302, 1998.

[17] Fullerton, D. and Walls, M., *Trash Talk*. RFF Weekly Policy Commentary, http://www.rff.org/rff/News/Weekly_Policy_Commentary/11_30_07_FullertonWalls_Commentary.cfm 2007.

[18] Coetzee, B. Personal Communication, City of Cape Town: Solid Waste Management Department, Cape Town, 2007.

[19] Huber, R., Ruitenbeek, J. and Seroa da Motta, R., *Market based instruments for environmental policymaking in Latin America and the Caribbean: Lessons from eleven countries*. Washington, DC. World Bank, 1997.

[20] O'Connor, D. and Turnham, D., *Managing the Environment in Developing Countries*. Policy Brief No 2. Paris. OECD Development Centre, 1992.

[21] Organisation for Economic Cooperation and Development, *Environmentally related taxes: Issues and strategies*. Policy brief. Organisation for economic cooperation and development, 2001.

Section 3
Cost benefits analysis

Integrated assessment of flood protection measures in the context of climate change: hydraulic modelling and economic approach

B. J. Dewals[1,2], E. Giron[3], J. Ernst[1], W. Hecq[3] & M. Pirotton[1]
[1]Department ArGEnCo, University of Liege, Belgium
[2]Fund for Scientific Research F.R.S.-FNRS, Belgium
[3]Centre for Economic and Social Studies on the Environment, Université Libre de Bruxelles, Belgium

Abstract

In order to protect our societies from damaging impacts of climate change (floods, heat waves, droughts etc), the most cost-effective adaptation measures must be selected among a wide range of options (including structural and non-structural protection measures). The Belgian national project "ADAPT" aims to provide guidance for this choice by developing a decision-support tool for the selection of protection measures against increased risk resulting from climate change, and more specifically from floods. This tool is based on a combination of cost-benefit analysis (CBA) and multicriteria analysis (MA), taking into consideration hydraulic, economic, social and ecological indicators. In order to demonstrate its efficiency, the method is applied in the two main Belgian river basins (Meuse and Scheldt).

Within this global framework, the present paper covers the description of the hydraulic modelling component and the economic approach, focusing on their interactions, for the case of the Meuse Basin (river Ourthe). The hydraulic simulations are performed with the two-dimensional numerical model WOLF 2D, which is developed at the University of Liege. High resolution Digital Elevation Models are exploited, enabling the representation of streets, buildings and parcels individually. In parallel, for past flood events, a relationship is established between simulated local water depths and actual damage costs, enabling the development and validation of suitable damage functions.

Keywords: climate change, hydrodynamic modelling, damage function, finite volume, digital elevation model, decision support.

WIT Transactions on Ecology and the Environment, Vol 108, © 2008 WIT Press
www.witpress.com, ISSN 1743-3541 (on-line)
doi:10.2495/EEIA080151

1 Introduction

Climate change is currently regarded as one of the most important threats to the environment and human well-being. This was confirmed recently by the release of the IPCC Working Group II Report on 'impacts, adaptation and vulnerability' (IPCC, April 2007) [12]. It is very likely that our societies will be confronted with increases in temperature, heat waves, heavy precipitation events, incidences of drought and rising sea level. If nothing is done to limit these threats, the phenomena will generate multiple damages that will put an increasing burden on our societies, as is demonstrated by the Stern Review (2006).

Policymakers can respond to these threats in two ways. First, by the development of *mitigation measures*, with the purpose of reducing GHG emissions, and secondly by the development of *adaptation measures* that aim to limit the impacts of climate change phenomena on populations and ecosystems.

The Belgian national research project "*ADAPT - Towards an integrated decision tool for adaptation measures*" focuses on inundation and on the development of corresponding *adaptation* measures. The project aims to meet the needs to support decision-making on adaptation policy by the development of an integrated decision-support tool for selecting adaptation measures. However, this development of adaptation policies is far from straightforward because of the uncertainties on the extent and the location of climate change impacts.

To develop the tool, the project focuses on the case study of flooding in Belgium. The development of the tool is based on an in-depth study of two case study areas that are located in the two main Belgian river basins (the Scheldt and the Meuse). Those study areas have been selected on the basis of their flood history, as well as their hydraulic, economic, social and ecological features.

By combining inundation maps and land use data, the developed methodology evaluates which elements or assets are affected by a given flooding scenario and how much they are affected in terms of inundation depth or other flooding characteristics [12]. The present paper provides first an overview of this procedure developed to combine hydrodynamic modelling results with the economic analysis of the impacts of flooding (section 2). Second, the hydrodynamic modelling (section 3) and the economic damage evaluation (section 0) are subsequently described in detail.

2 Overview of the integration between flood modelling and evaluation of economic damage

As shown in Figure 1, the evaluation of the risk associated to floods relies basically on two main parameters: the *probability* of occurrence of a flood, which can be related to the statistically calculated return period of the corresponding discharge, and its *consequences*. Consequently, the *risk* can be evaluated mathematically by the product of the probability times the consequences, summed over each individual event.

The probability of occurrence of a flood event is ascertained by statistical hydrological methods that assigns a *return period* (in years) to a given discharge.

FLOOD RISK = sum (probability * consequences)			
Probability	Consequences (actual predicted damage)		
Probability (return period)	**Exposure** (extent, depth, velocity)	**Elements-at-risk** (people, buildings, roads, networks, eco-systems ...)	**Vulnerability** (susceptibility, adaptative capacity, resilience, ...)

Figure 1: Main components of flood risk evaluation.

The major components to be considered for quantifying the consequences are the *exposure*, the *elements-at-risk* and their *vulnerability*.

The exposure characterizes the inundation by means of the three following features: inundation extent, water depth and flow velocity. This information is directly provided for instance by a 2D hydrodynamic modelling of the flood. Additional information may also be relevant, such as flood duration or water quality indicators.

Second, the elements-at-risk must be clearly identified. Indeed, flood damages would obviously remain extremely low if the inundated area, whatever its extent, contains no element of socio-economic (or ecologic) importance. For instance, for given flow conditions, the "flood risk" would vary significantly depending on whether the inundation takes place in a densely populated area or in a rural one...

For large scale studies (e.g. regional or national), the elements-at-risk may be estimated by a percentage of the total studied area [17], while for more detailed analysis, they are identified individually by means of land use database.

Finally, the vulnerability of the elements-at-risk (depending on their susceptibility, adaptative capacity and resilience) is introduced in the risk analysis by means, for instance, of *damage functions* (section 0).

In the present study, the consequences are preliminarily restricted to the economic damage, in such a way that the flood risk can be quantified by a monetary value (€/year). Therefore, economic damage functions are exploited. They provide an estimation of the *relative damage* for each identified element-at-risk, based on the inundation characteristics, which are the outcomes of the 2D hydrodynamic modelling. Although the flow velocity and the flood duration [10] may play a significant part for specific applications (e.g. dam break flows) and should be included in the analysis for such cases, the water depth remains the most influential parameter for a vast majority of applications.

The relative damage represents the percentage of the total value of goods that is lost as a result of the flood event. Consequently, combined with the monetary value of the elements-at-risk, the relative damage may be translated into an absolute economic loss.

Such a comprehensive analysis enables to compare different scenarios of flooding, possibly taking into consideration effects such as climate change or flood protection measures, and to communicate more effectively about the outcomes of such studies, notably because the conclusions may be expressed in terms of monetary values instead of more technical hydraulic parameters.

3 Hydrodynamic modelling

In order to validate the combined hydraulic-economic evaluation of the impacts of flooding, the methodology is first tested and refined for a case study area, which covers two reaches of the river Ourthe in the Meuse basin (Belgium, Figure 2). Besides, prior to evaluating the benefits of different adaptation measures, a preliminary step consists in determining how inundation hazard and its economic impact are likely to be affected by climate change. More precisely, for several return periods of floods and for a number of climate change scenarios, two-dimensional hydrodynamic simulations have been run in order to identify which modifications climate change will cause to inundation extents as well as to water depths and flow velocities in the floodplains.

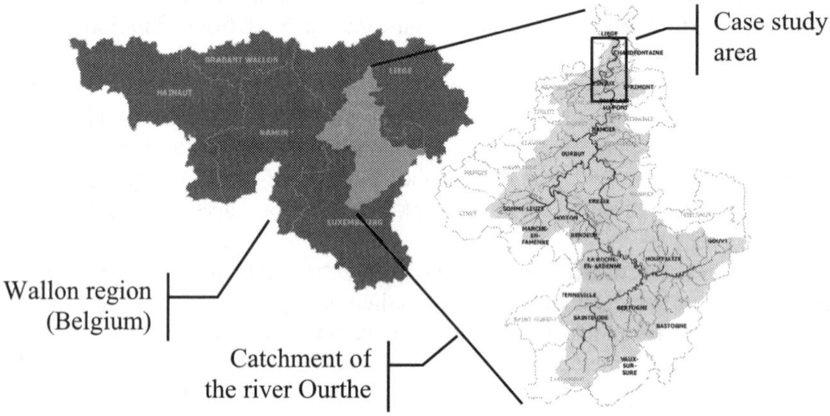

Figure 2: Location of the case study area in the catchment of the river Ourthe.

3.1 Hydrodynamic model WOLF 2D

The HACH research unit from the University of Liege develops a wide range of numerical tools for simulating free surface flows and transport phenomena. Those computational models are interconnected and integrated within one single *modelling system* named *WOLF*. Among those computation units, the two-dimensional flow model *WOLF 2D* is exploited for the present study and is described hereafter. WOLF 2D is based on the depth-averaged equations of volume and momentum conservation, i.e. the "shallow-water" equations [2]:

$$\frac{\partial \vec{s}}{\partial t}+\frac{\partial \vec{f}}{\partial x}+\frac{\partial \vec{g}}{\partial y}+\frac{\partial \vec{f}_{\mathrm{d}}}{\partial x}+\frac{\partial \vec{g}_{\mathrm{d}}}{\partial y}=\vec{S}_0-\vec{S}_{\mathrm{f}}, \tag{1}$$

with $\vec{s}=\begin{bmatrix} h & hu & hv \end{bmatrix}^{\mathrm{T}}$ the vector of the conservative unknowns. $\vec{S}_0$ and $\vec{S}_{\mathrm{f}}$ designates respectively the bottom slope term and the friction term:

$$\vec{S}_0 = -gh\left[0 \quad \partial z_b/\partial x \quad \partial z_b/\partial y\right]^T, \qquad \vec{S}_f = \left[0 \quad \tau_{b,x}/\rho \quad \tau_{b,y}/\rho\right]^T. \qquad (2)$$

$\vec{f}$ and $\vec{g}$ represent the advective and pressure fluxes in directions x and y, while $\vec{f}_d$ and $\vec{g}_d$ are the diffusive fluxes.

Written out in full, these fluxes become:

$$\vec{f} = \begin{pmatrix} hu \\ hu^2 + \frac{1}{2}gh^2 \\ huv \end{pmatrix}, \quad \vec{f}_d = -\frac{h}{\rho}\begin{pmatrix} 0 \\ \sigma_x \\ \tau_{xy} \end{pmatrix}, \quad \vec{g} = \begin{pmatrix} hv \\ huv \\ hv^2 + \frac{1}{2}gh^2 \end{pmatrix}, \quad \vec{g}_d = -\frac{h}{\rho}\begin{pmatrix} 0 \\ \tau_{xy} \\ \sigma_y \end{pmatrix}. \qquad (3)$$

The following notations have been used: t represents the time, x and y the space coordinates, h the water depth, u and v the depth-averaged velocity components, z_b the bottom elevation, g the gravity acceleration, ρ the density of water, τ_{bx} and τ_{by} the bottom shear stresses, σ_x and σ_y the turbulent normal stresses, and τ_{xy} the turbulent shear stress. The bottom friction is conventionally modelled thanks to an empirical law, such as the Manning formula. The models enable the definition of a spatially distributed roughness coefficient. The internal friction may be reproduced by different turbulence closures included in the modelling system, such as simple algebraic ones but also a complete depth-averaged k-ε model.

The numerical model deals with multiblock Cartesian grids. This feature increases the size of possible simulation domains and enables local mesh refinements, while preserving the lower computation cost required by Cartesian grids compared to unstructured ones. A grid adaptation technique restricts the simulation domain to the wet cells. Besides, wetting and drying of cells is handled free of volume conservation error [11].

The space discretization is performed by means of a finite volume scheme. Variable reconstruction at cells interfaces is performed linearly, in combination with slope limiting, leading to a second-order spatial accuracy. The advective fluxes are computed by a Flux Vector Splitting (FVS) technique developed by the first and fifth authors. A Von Neumann stability analysis has demonstrated the stability of this FVS [4]. The diffusive fluxes are legitimately evaluated by means of a centred scheme. Since the model is applied to compute steady-state solutions, the time integration is performed by means of a 3-step first order accurate Runge-Kutta algorithm, providing adequate dissipation in time. The time step is constrained by the Courant-Friedrichs-Levy (CFL) condition based on gravity waves. A semi-implicit treatment of the bottom friction term (2) is used, without requiring additional computational costs.

The model has been extensively validated and has shown its efficiency for numerous practical applications [4, 5, 8, 9, 11].

3.2 Main assumptions and considered return periods

Results of Global Circulation Models (GCM) and Regional Climate Models (RCM) provide estimates of the potential increase in precipitation (winter and summer) and potential changes in evapotranspiration as a result of climate change [1]. Those predicted changes are affected by a significant level of uncertainty due to the climate models themselves and, to an even greater extent, to the discrepancies in the scenarios used for running those climate models [13].

Moreover, translating those changes in precipitation and evapotranspiration into changes in river discharges is also a challenging task, since the catchment response depends on many factors such as the land roughness and permeability, the rainfall duration, … Therefore, at the present stage of the ADAPT project, simple assumptions have been considered regarding the expected changes in the peak discharges of the river Ourthe as a result of climate change (CC). Indeed, according to a comprehensive literature review [3, 6, 7], an increase by 10% of flood discharge may be regarded as reasonable. Moreover, in order to evaluate the sensitivity of the results with respect to this perturbation factor on the discharge, increases by 5% and by 15% have also been considered. Finally, a more "extreme" case has been simulated as well (+30%). All those assumptions will eventually be confirmed and refined by comparison with the output of a hydrological model [3]. The study is performed for two different return periods, namely 25 and 100 years, for which the best adaptation strategies will be complementary, due to the significant differences in inundated areas and in the probability of occurrence.

A high resolution Digital Elevation Model (LIDAR data) is used, enabling thus to represent each street or building individually on a 2m × 2m computation grid.

3.3 Simulation results

By means of hydrodynamic modelling, the assumed modifications in the discharge values have been translated into updated evaluations of inundation extent and characteristics for the considered reaches of the river Ourthe. Table 1 summarizes the different hydraulic simulations carried out. It also provides averaged values of the change in water depths for CC scenarios 1 to 3.

Table 1: Mean change in water depth induced by Climate Change (CC) for two different return periods (25-year and 100-year floods).

		Base scenario	*Scenario 1 (+5%)*	*Scenario 2 (+10%)*	*Scenario 3 (+15%)*
25-year flood	Discharge	726 m³/s	799 m³/s	835 m³/s	835 m³/s
	Reach n°1	-	+25cm	+40cm	+40cm
	Reach n°2	-	+25cm	+40cm	+40cm
100-year flood	Discharge	876 m³/s	964 m³/s	1007 m³/s	1007 m³/s
	Reach n°1	-	+30cm	+45cm	+45cm
	Reach n°2	-	+40cm	+60cm	+60cm

Results have also been produced in the form of updated inundation maps, indicating, for each return period and each climate change scenario, the 2D distribution of water depth in the floodplains, flow velocity and increase in water depth compared to the base scenario. It is found that the results can be very sensitive to the perturbation factor affecting the discharge, due for instance to the reduced efficiency of flood protection structures (e.g. dikes) above a threshold value of discharge (design discharge). Moreover, the complexity of the flow fields represented on the inundation maps recalls the relevance of exploiting a fully two-dimensional flow model. The present set of hydrodynamic modelling results serves as an input for the subsequent economic damage modelling, but also for the evaluation of social and ecologic impacts.

4 Economic damage modelling

This section covers an in-depth description of the methodology developed for evaluating the economic damage induced by floods, while a short overview of the procedure was introduced in section 2.

4.1 Geographic database

Accurate land use data constitute key input data in any analysis of the socio-economic impacts of floods. In the present study, a database from the Belgian *Institut Géographique National* ("Top10v-GIS" - © IGN) is exploited. It gathers very comprehensive sets of vector data, contained in information files formatted as ESRI *ShapeFiles*, which can be imported within most GIS environments.
The database contains 18 layers of data, each one referring to a class of objects (e.g. administrative data, altimetry, electricity, land use, structure …). The scale of those maps corresponds to 1:10000 and their global accuracy is 1 meter in the Lambert Belgian reference system. In this study, five different types of objects in land use and structures layers are taken into consideration: residential and industrial buildings, road networks, crops and fields, as well as forests.

4.2 Exploitation of hydrodynamic modelling results in combination with vector land use database

As shown in Figure 3, the estimation of the absolute economic damage may be described as a four-step procedure:

- first, the inundation extent is extracted from the inundation maps and is exploited in combination with land use data to identify the elements-at-risk;
- second, a specific damage function is assigned to each of the five classes of elements-at-risk, representing their respective vulnerability to flooding;
- next, this function is applied, in combination with the water depths (and possibly flow velocity or flood duration), to provide a relative value of the damage encountered by each object during the considered flooding event;
- finally, the relative damage is multiplied by an estimated value of the flooded element (so called: *specific patrimonial value)*. At the end of this methodology, an absolute estimation of the financial damage induced by the flood event can be assigned to each node of the initial flood map grid.

For the case study, housing values will be collected from local cadastral offices.

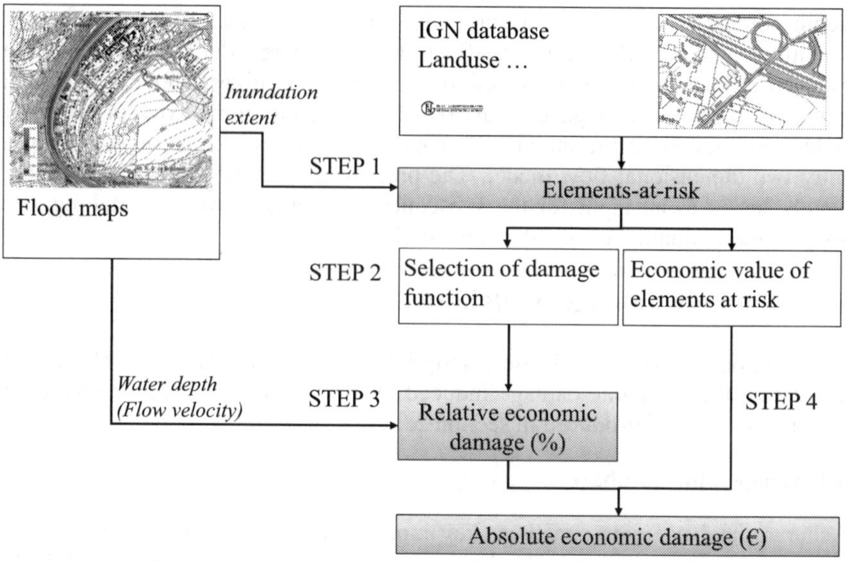

Figure 3: Flow chart describing the integration of hydrodynamic modelling results and economic damage evaluation.

It must be noted that most damage functions used for economic loss evaluations as a result of flooding depend on water depth, which is a direct outcome of the hydrodynamic simulations. However, because these numerical simulations are not run on a *Digital Terrestrial Model* (DTM) but on a *Digital Surface Model* (DSM, including anthropic over grounded objects), the value of the water depth at the location of buildings is not directly available in the output of the numerical modelling. Therefore, these water depths are calculated by means of an average of the water depth values in the vicinity of the building and are then assigned to each computation cell representing the corresponding over grounded object.

4.3 Damage functions

The damage functions, linking the hydraulic parameters and the type of elements-at-risk to the damage costs, can take various forms: continuous function, scale function, simple ratios, to be applied to the value of the good affected or providing a direct monetary cost of the damages … [15]. These functions are developed either by feedback from real events (observation of damages declared after flooding), or by expert judgment (development on the basis of the experience of individuals carrying out the appraisal of loss or from building experts). Experience of the latest flooding events in Belgium shows that damages on houses are the most important contribution to the total damage compared to other sources of damages. Consistently, the present development of damage functions is focused mainly on this type of damage.

The damage functions generally link the following two parameters:

- "*water level*", i.e. water depth inside the considered object (or good), measured from the floor of the habitation;
- relative damage, expressed as a percentage of the value of the affected object.

The following damage function reported in the ICPR Rhine atlas [14] is presently used: $D = 2\,H^2 + 2\,H$, where H is the water level recorded during the flooding and D designates the damage in percentage of the housing value.

These functions are obtained by empirical observation of the losses and statistical processing of collected data. In other approaches, the calculation of damages is carried out directly by means of *absolute* damage functions [12]. The value of the assets is already integrated in the damage functions, such as for instance those provided by the *Multi Coloured Manual* in the UK [16].

4.4 Development and validation of damage functions

The methodology of the development and validation of these damage functions in the present study follows 3 steps: *(i)* a literature review, *(ii)* the application of a number of damage functions based on the results of hydrodynamic modelling for past flood events on the case study of the river Ourthe and *(iii)* the comparison of computed damage with reference values collected by the Belgian "*Disaster Fund*" after real flood events. Available data of real damages, per street and per category, contain on one hand values *claimed* by the population and on the other hand values eventually *accepted* to be supported by the "Disaster Fund". These two values differ by a factor of the order of two.

Partly due to a lack of information in the reference data collected by the Disaster Fund, computed values of damage appear to be globally overestimated according to the present preliminary comparisons. Nevertheless, as shown in Figure 4, the economic damage modeling procedure succeeds in providing a satisfactory overall picture of the relative cost of damage caused by the major recent flooding events along the river Ourthe, namely the 1993, 1995, 2002 and 2003 floods.

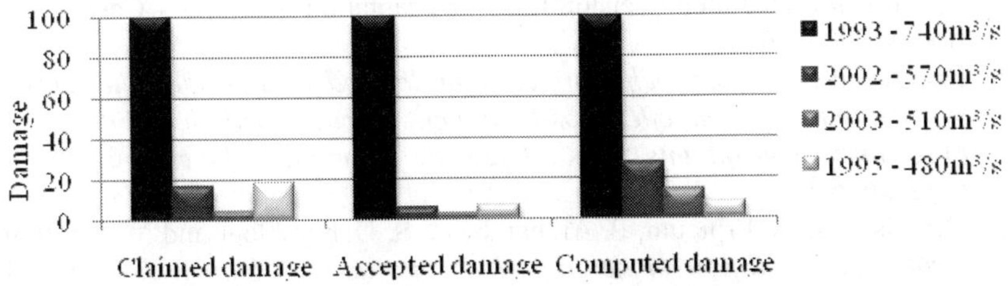

Figure 4: Damage claimed to / accepted by the Disaster Fund vs computed damage, expressed in percentage of the corresponding damage of the 1993-flood.

5 Conclusion

The present paper describes the combination of two-dimensional hydrodynamic modelling with a procedure for economic damage evaluation in the context of climate change. The high resolution topographic data used for the flow simulations lead to very detailed flood maps, which serve as an input for computing the cost of damage. This evaluation takes into consideration the three following components: a detailed vector land use database, suitable damage functions and specific values of the elements-at-risk. The overall procedure leads to satisfactory estimations of the relative damage caused by a series of real past major flood events along the river Ourthe (Meuse Basin, Belgium).

Acknowledgements

Part of this research was carried out on behalf of the Belgian Science Policy (BELSPO), in the framework of the program "Science for a Sustainable Development". The authors also gratefully acknowledge the Walloon Ministry of Facilities and Transport (MET) for the Digital Elevation Model and other data.

References

[1] Boukhris, O., P. Baguis, P.Willems and E.Roulin, *Climate change impact on hydrological extremes along rivers and urban drainage systems. II.Study of climate change scenarios*. May 2007, K.U.Leuven - Hydraulics Section & Royal Meteorological Institute of Belgium: 92 p.

[2] Chaudhry, MH, *Open-Channel Flow*. 1993, Englewood Cliffs: Prentice Hall, 483 p.

[3] De Groof, A., W. Hecq, I. Coninx, K. Bachus, B. Dewals, M. Pirotton, M. El Kahloun, P. Meire, L. De Smet and R. De Sutter, *General study and evaluation of potential impacts of climate change in Belgium*. ADAPT - Towards an integrated decision tool for adaptation measures - Case study: floods. 2006, 66 p.

[4] Dewals, B., *Une approche unifiée pour la modélisation d'écoulements à surface libre, de leur effet érosif sur une structure et de leur interaction avec divers constituants*. 2006, Thèse de doctorat, Université de Liège, Liège: 636 p.

[5] Dewals, B.J., S. Erpicum, P. Archambeau, S. Detrembleur and M. Pirotton, *Depth-integrated flow modelling taking into account bottom curvature*. J. Hydraul. Res., 2006. **44**(6): p. 787–795.

[6] Dewals, B.J., P. Archambeau, S. Erpicum, S. Detrembleur and M. Pirotton. *Identification d'impacts locaux du changement climatique ét évaluation de mesures d'adaptation grâce aux modèles hydrologiques et hydrodynamiques Wolf*. in *Variations climatiques et hydrologie*. 2007. Lyon, France: SHF.

[7] Dewals, B.J., R. De Sutter, L. De Smet and M. Pirotton, *Synthesis of primary impacts of climate change in Belgium, as an onset to the*

development of an assessment tool for adaptation measures. Geophysical Research Abstracts, 2007. **9**(11217): p. 2.

[8] Dewals, B.J., S.A. Kantoush, S. Erpicum, M. Pirotton and A.J. Schleiss, *Experimental and numerical analysis of flow instabilities in rectangular shallow basins.* Environ. Fluid Mech., 2008 (accepted).

[9] Dewals, B.J., S. Erpicum, P. Archambeau, S. Detrembleur, M. Pirotton, *Numerical tools for dam break risk assessment: validation and application to a large complex of dams*, in *Improvements in reservoir construction, operation and maintenance*, H. Hewlett (ed). 2006, Thomas Telford: London. p. 272–282.

[10] Dushmanta, D., H. Srikantha and M. Katumi, *A mathematical model for flood loss estimation.* J. Hydrol., 2003. **277**: p. 24–49.

[11] Erpicum, S., *Optimisation objective de paramètres en écoulements turbulents à surface libre sur maillage multibloc.* 2006, PhD thesis, University of Liege: 356 p.

[12] FLOODSITE, *Guidelines for socio-economic flood damage evaluation.* 2006.

[13] Intergovernmental Panel on Climate Change, *Climate change 2007: Synthesis Report - Summary for policymakers.*

[14] International Commission for the Protection of the Rhine, *ICPR Rhine atlas*. 2001.

[15] Ledoux consultants, Cereve and Cemagref, *Synthèse des évaluations socio-économiques des instruments de prévention des inondations*, Rapport au Ministère de l'écologie et du développement durable.

[16] Penning-Rowsell, E.C., C. Johnson, S.M. Tunstall, S.M. Tapsell, J. Morris, J.B. Chatterton, A. Coker and C. Green, *The Benefits of Flood and Coastal Defence: Techniques and Data for 2003*, Flood Hazard Research Centre, Middlesex University.

[17] Van der Sande, C.J., S.M. de Jong and A.P.J. de Roo, *A segmentation and classification approach of IKONOS-2 imagery for land cover mapping to assist flood risk and flood damage assessment.* International Journal of Applied Earth Observation and Geoinformation, 2003. **4**: p. 217–229.

A projection of BIPV systems in Taiwan: costs versus benefits

K.-J. Hsu
Department of Construction Technology, Leader University, Taiwan

Abstract

Because Photovoltaic system tends to save the energy expenditure across the life span of the systems, this paper uses life cycle costing (LCC), saving-to-investment ratio (SIR), and discounted payback period (DPP) methods to construct the renewable energy evaluation framework in Taiwan. The benefits of PVs mainly derive from energy cost savings, enhanced power quality and reliability, reduced environmental emissions, rebates, and other incentives. Based on the field data of PV-projects located at Taiwan, the paper identifies the technical and economic parameters of the systems evaluation model. The empirical case showed that the saving-to-investment ratio of the BIPV systems with/without government subsidy is not currently feasible in Taiwan. After examining critical factors of the PVs-project, the sensitivity analysis of the future PVs-project was shown. Based on the empirical residential case with a reasonable PR value, the results of sensibility analysis showed that BIPV will become one of the dominant renewable energies in the near future if good, integrated building design is available.

Keywords: building-integrated photovoltaic (BIPV), building economics, life cycle costing, saving-to-investment ratio, risk analysis.

1 Introduction

Photovoltaic systems (PVs) offer clean, virtually silent and low maintenance power on site. Its potential application depends on the local social-economic development and environmental conditions. It was recently suggested [1–3] that building Integrated Photovoltaic (BIPV) systems consisting of integrated photovoltaic modules into the building envelope could be a cost-effective means of abating CO_2 emissions. There has been a dramatic down turn in costs [1, 4, 5],

WIT Transactions on Ecology and the Environment, Vol 108, © 2008 WIT Press
www.witpress.com, ISSN 1743-3541 (on-line)
doi:10.2495/EEIA080161

for example, in the 1970s PV modules cost in the region of £15,000/kWp, but in the 1990s costs were down to less than £3,000/kWp [4].

BIPV systems are increasingly used in building design around the world and have become the symbol and language of innovative, high-tech eco-building. However, one of the major barriers is still the issue of financial feasibility. Hsu et al. [6] used life cycle cost analysis and saving-to-investment ratio to analyze the empirical data of a residential BIPV project located at Kao-Hsiung. The results showed that compared to low electricity cost on the Island, it was still not economically feasible for a home owner/occupant, even with the government subsidy.

Taiwan is not a member of IEA, however, concerning the renewable energy issue, the Bureau of Energy Ministry of Economic Affairs has proposed a subsidized program for those PV system users. More PVs-projects will be promoted in the near future, but in the long run the major barrier is whether a PV-project makes economic sense. This paper looks at the benefit in the reduction of carbon dioxide emissions against the economic evaluation of the BIPV systems. How the key variables will affect the feasibility of the system now and in the future will be discussed.

Instead of focussing on the cost-effective analysis as in previous research, this paper is more concerned with the break-even point of PVs investment. The direct financial benefit of a BIPV system is primarily the energy generated; so in order to examine the effect of discount rate and energy price escalation rate on discounted payback years, the sensitivity analysis was done. On the other hand, because high-humidity climate will affect the efficiency of the PVs in Taiwan, how a ventilated BIPV design will affect the payback years will also be addressed. After examining the performance ratio in Taiwan, the PVs cost down effect and ventilated-BIPV technical improvement will be discussed.

In the following section, the economic evaluation model of the PV system will be developed first. Section 3 introduces the cases that use renewable energy and expounding the economic meaning of the LCC and SIR analysis. Then the sensibility analysis based on DPP method will be extended to discuss the future of PV systems in Taiwan. Finally, the conclusion and discussions were drawn.

2 The model

In life cycle cost (LCC) analysis, we need to consider the difference between alternatives when two or more projects are being compared. When evaluating a PVs-project, energy savings need be considered, costs that are common among alternatives that do not affect the selection decision, can be ignored. Assume the discount rates among the discounting periods are known with certainty and invariant, and compare the cost-effectiveness among different kinds of energy alternatives. A comparison between the present values of life cycle cost (*PVLCC*) of the renewable energy system to the traditional system determines whether it is cost effective. If the PVLCC is lower than for the traditional base case and in other aspects is equal, and the project meets the investor's objectives and budget constraints, it is considered cost effective and the preferred investment [7–9].

On the other hand, because of energy savings and the carbon dioxide (CO_2), emissions abatement is the major benefit sources of a PVs-project. The present value of a PVs-project can be written as

$$PW = -C - M\left[\frac{(1+i)^n - 1}{i(1+i)^n}\right] - \sum_{j=5,10,15,..}\left[\frac{RP_j}{(1+i)^j}\right] + R_0\sum_{j=1}^{n}\left(\frac{1+e}{1+i}\right)^j . \quad (1)$$

Where PW represents present worth of the cost of the project over the project life of n years; *C* represents initial cost of the facility. $M\left[\frac{(1+i)^n - 1}{i(1+i)^n}\right]$ represents the present value of the operation and maintenance cost (*PVM*), and *i* represents discount rate. $\sum_{j=5,10,15,..}\left[\frac{RP_j}{(1+i)^j}\right]$ represents the present value of the repair or replacement fee (RP_j) at year *j* of the facilities (*PVRP*). $R_0\sum_{j=1}^{n}\left(\frac{1+e}{1+i}\right)^j$ represents the present value of energy saving respect to year *j* in the facilities (*PVR*).

When we evaluate the energy savings of the project, because the price of energy has the characteristic of changing constantly, the energy shortage may cause the violent fluctuation of the price of energy in the future. Considering the rate of energy price escalation trend (*e*) is always faster than the rate of general price inflation in Taiwan, the present value of energy saving can be evaluated using the following equation [10],

$$\sum_{j=1}^{n}\frac{R_j}{(1+i)^j} = \begin{cases} R_0\dfrac{(f^n - 1)f}{f-1}, & if\ f \neq 1 \\ nR_0 & , if\ f = 1. \end{cases} \quad (2)$$

In which $f = \frac{1+e}{1+i}$, R_j represents the net cash inflow from energy saving at year *j*, and R_0 is the energy saving benefit respect to initial energy price. For example, if the annual output of the PVs-project is *E*, and the initial unit price of the traditional power is P_0, the annual energy savings at initial time are $R_0 = EP_0$. Because *PVM*, *PVRP* are only a small part of the PVLCC, the pay-back year of the project can be simplified as follows:

$$PP_{DCF} = \frac{ln\left[1 + \left(\frac{C + PVM + PVRP}{R_0}\right)(f-1)\right]}{ln\,f}. \quad (3)$$

Discounted payback period (DPP) method calculates the time point at which accumulated savings net of the accumulated cost is sufficient to offset the initial costs. The payback year method is essentially a break-even measure which can be used to determine the minimum time a BIPV system must last to recover the investment costs. The DPP is often used as a rough guide to cost-effectiveness. Extending the results to the future prospect discussion, the sensitivity analysis of the DPP is quite a suitable tool.

3 The critical factors of economic feasible for PV systems

The parameters of a PVs-project feasibility analysis include: the financial parameters (discount rate, price escalation rate), and the technical parameters of PVs (the life span of PVs, timing of component replacement, cost of initial cost, etc.). Treating PVs as consumable goods for the household, the opportunity cost of the consumable goods can be estimated as the average fixed deposit rate of Taiwan. Considering the financing behaviour characteristic of the general family in Taiwan, in this paper we set the discount rate as 3%. The financial parameters in this paper mostly refer to statistical data from the Directorate General of Budget, Accounting and Statistics, Taiwan. The technical parameters mostly refer to datasheets from the respective manufacturer, for example, the life of the PV panel is assumed as 20 years.

The results show that the present value of the opportunity cost larger than the present value of the energy savings from systems. The net present value of the system is negative. Without the government subsidy, the SIR of the residential off-grid case located at Kao-Hsiung is about 13.7% and with the government subsidy about 21.9%. In case of grid-connected industrial facility located at Taichung the SIR upgrade to 16.6% in respect to the case without subsidy, and 26.8% in respect to the case with subsidy. Low energy saving benefit and high technological cost are the major reasons that result in the PV systems not being economically feasible in Taiwan now.

The PVs-project output is affected by the solar radiation amount available on the site and the performance ratio (PR) of the systems. Given an average PR value evaluated at yearly basis and the local solar radiation available at the site (dimensionless), the annual output (kWh) of the PVs project can be calculated as follows:

$$E = PR \bullet Y_r \bullet W \,. \tag{4}$$

Let the reference yield equal to total in-plane irradiance divided by the PV's reference irradiance ($Y_r = H / G$, hour); W represents the nameplate d.c. power of the installed PVs (kW). If G equals 1 kWh/m^2, then Y_r is the number of peak sun-hours, it represents an equivalent number of hours at the reference irradiance. By normalizing with respect to irradiance, PR value quantifies the overall effect of losses on the rated output. Because of losses due to PV module temperature, PR values are greater in the winter than in the summer [11].

In the following subsections, the key factors examined are (a) solar radiation, (b) PR, (c) thermal and ventilated-BIPV design issues, (d) energy price and its escalation trend which affect savings of the systems.

3.1 Solar radiation in hot-humidity region

The level of solar resource has an impact on the output of a PVs-project. Taiwan located at low latitude about 23°N, is hot-humidity region with slight differences of temperature and slight changes of solar radiation in the four seasons. However, unstable climate such as cloud and rain in the spring and typhoons in the summer, lower the annual solar radiation. Compared to bell shape solar radiation across the year in low-high latitude region, the daily radiation across the year of Taiwan has only a slight change.

Table 1 shows that the variation of annual solar radiation across the island ranges from 801 to 1560 Wh/m^2/year, and the average is lower than 48°00'N at the Freiburg. Figure 1 shows the daily output of "stand-alone type" BIPV systems in Southern Taiwan, it seems there is little significant difference across the whole year. Most PVs research papers concluded that sites closer to the equator have higher solar resources [2, 12]. However, Taiwan is an unusual case, as the micro-climate patterns affect the PVs output to a greater degree.

Table 1: Annual radiation comparison across different latitude [12, 13].

	Latitude	Annual solar radiation (Wh/m^2/year)
Taiwan-1 (Max)		1,560
Taiwan-2 (Average)	±23°N	1,059
Taiwan-3 (Min)		801
Sahara	25°00'N	2,500
Israel	33°00'N	2,000
Los Angles, USA	34°30'N	1,816
Athens, G	37°58'N	1,580
Trapani, I	38°00'N	1,800
New York, USA	40°47'N	1,424
Rome, I	41°48'N	1,529
Toronto, C	43°40'N	1,380
Freiburg, D	48°00'N	1,100
London, UK	51°29'N	898
Berlin, G	52°28'N	1,026
Helsinki, FIN	60°00'N	950
Lerwick, UK	60°00'N	775

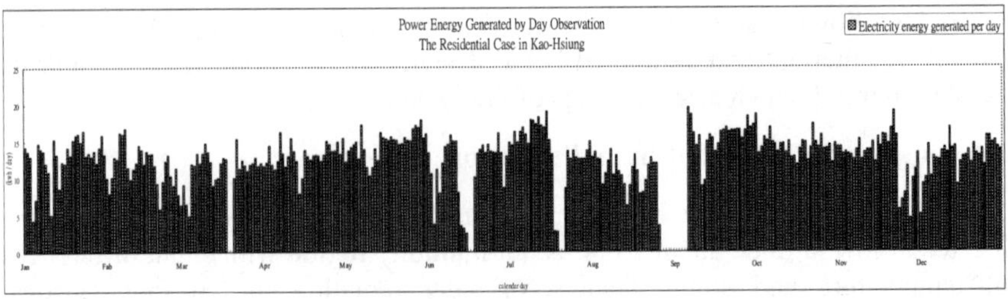

Figure 1: Daily power output across year of the residential off-grid BIPV-project located at Kao-Hsiung.

3.2 Performance ratio

The productivity of PVs-project relative to the solar radiation it receives is measured by the PR value, although there are some special grid-connected cases demonstrated by government showing that PR value is greater than 80%. In the empirical case of "stand-alone type PVs-project", system output will be shut-off after the battery is fully charged. According to field data of the cases study here, the residential roofing PVs showed that the PR value 53.2%, is lower than the grid-connected case of a Taichung Industrial facility 65.5%.

The low PR value in the residential case located at Kao-Hsiung can be traced to the use of a stand-alone system. This is because main electricity systems do not provide incentive and good connected network support for individual small micro-generations. Taiwan is a country at the earliest stage of starting to use PVs. Most PVs PR values in Taiwan are lower than in London, New York, Los Angeles and Germany, where the average PR is 70% [2, 11, 13].

3.3 Thermal issue and ventilated-BIPV

As BIPV modules are installed on building envelope, obviously the changes of exterior environment or climatic conditions will affect their total performance, hot-humidity climate in Taiwan results in high temperature, and low PR value. Among a variety of factors, heat has a direct influence on PV. If the temperature of PV is above 25°C, the effect of temperature arise on solar cell efficiency is negative and material dependent. For example, the efficiency of crystalline solar cells falls whenever the temperature increases, the decrease in the rate of the efficiency is estimated at about 0.4%, which is lower than the rate under STC [12]. In contrast, the behaviour with diffused light, increased temperature, and shading for amorphous showed a high energy yield in kWh per kW_p and thus a high performance ratio.

An experiment of ventilated BIPV using CFD simulation showed that compared to traditional BIPV design, a ventilated BIPV saves about 8–26% of energy consumption. Also a ventilated BIPV can provide low solar cell temperature and good performance for PV systems. In order to ensure the nominal PV efficiency, the choice of solar cell material and appropriate ventilation are therefore necessary. In other words, the application of PVs installed in buildings must pay attention to ventilation and thermal issues, so as to have a better energy savings and to maintain a high PR value.

3.4 The benefits from energy savings and price escalation rate

The benefits of renewable energy are mainly derived from the saving of traditional energy and clean energy environment benefits. With LCC, we learn that the benefits of renewable energy rely on the pricing model of traditional energy. To boost their economies, developing countries tend to lower the cost of traditional energy to supply low-priced energy to industries. In Taiwan, the power supply is controlled by a company owned by the MOEA, a government-owned company. The price of energy tends to be low for overall economic development. The price does not reflect the real economic cost. This results in the benefits from the renewable energy project being relatively low.

The data shows that the past upward rate of the price of electricity lie between 1 to 2% in Taiwan, and the upward rate of the price of electricity is estimated by government to be about 2% per year. Because LCC is considered with the whole life cycle, in a situation that the petrochemical energy is limited, the relevant price of electricity, coal and gas certainly will be going up constantly in the future. Combining the price escalation rate of the electricity energy and using the observation of electricity energy generated in the first year as the estimation of the following lifetime energy savings, the illustrative example in the following section will analyze the sensitivity of DPP change with respect to energy price escalation rate change.

4 Sensitivity analysis

According to the results of LCC and SIR analysis, it is not economically feasible for a home owner-occupant of PVs in Taiwan. The result also shows that it cannot break-even in the life span of the system. The discounted payback years are larger than the life of the system. The reasons include thermal issue of BIPV design in the hot and humid climate region, energy cost and price policy, sell-back-price of the renewable energy, higher technology cost and lower electricity generation of PVs.

Comparing the results and economy of PVs to those of developed countries, the PVs cost in Taiwan is relatively high. On the other hand, the energy savings from PVs output in Taiwan is relatively low compared with Japan and EU countries. Considering the cost down trend of PVs and PV industry developed recently in the Island, the data used now represents the developing technologies of PVs rapidly becomes obsolete. The effects of cost trend have been analyzed in many papers [1–5].

The reason for low energy price include the land pipeline cost is lower for Tai-Power Company, and environmental cost is almost ignored. Therefore, Tai-Power has lower cost of power generation, which leads to lower power saving economic benefits. But in the progresses of economic and environmental development, the energy price policy has been keenly discussed. The high price energy and high escalation rate will be the critical issue the Island needs to face.

Critical parameters that influence the assessment result include low energy savings, cost down trend, study period, and discount rate etc. To meet the economic sense, the paper uses equation (3) and based case of the future prospect to analyze the sensitivity of energy price escalation rate and discount rate on DPP without subsidy. The base case conditions of the sensitivity analysis include EU documents showing that PV cost will be anticipated to decrease by half in 2010–2015; and based on the empirical residential case analysis located at Kao-Hsiung, by which enhanced PR value to 70%, an average level of a developed country.

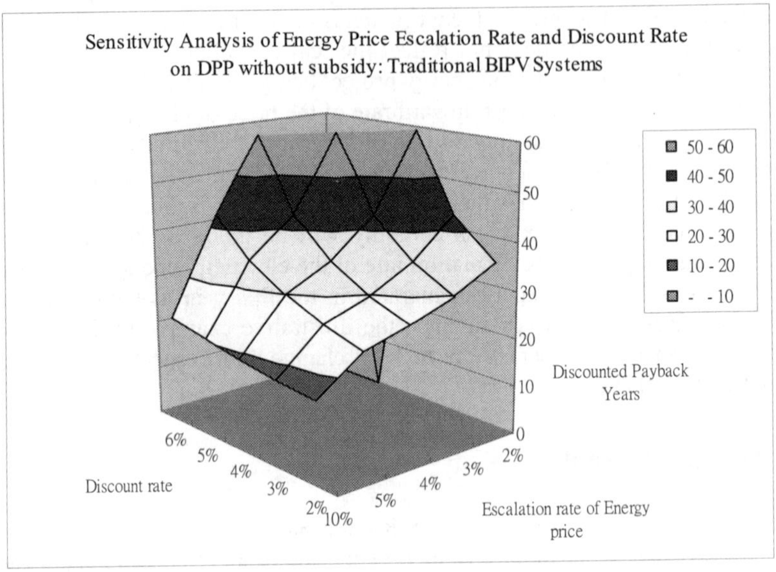

Figure 2: Sensitivity analysis of energy price escalation rate and discount rate on DPP without subsidy: traditional BIPV systems.

Fig. 2 shows the results of sensibility analysis for traditional BIPV respect to energy price escalation rate and discount rate on DPP. The Figure shows that with higher escalation rate of electricity price and lower discount rate, discounted payback years of BIPV systems will be less than 20 years. For example, when escalation rate of electricity price is 8% (this is most likely the energy market rate now) and discount rate is 3%, the DPP of the BIPV project will be down to 20 years. Higher escalation rate of electricity price will lower the DPP. Adopting ventilated-BIPV will lower the DPP threshold more.

Fig. 3 shows the simulation results in the case of ventilated-BIPV. The Figure shows that Ventilated BIPV can achieve shorter payback years. When escalation rate of electricity price is 8% and discount rate is 3%, the DPP of the BIPV project is down to 18 years. This should be feasible in a real case, and is thus good for an architect making energy-saving designs. Under such a case, BIPV will become one of the dominant renewable energies in the near future if integrating good building design is available. It is worth studying further and exploring the potentials in advance.

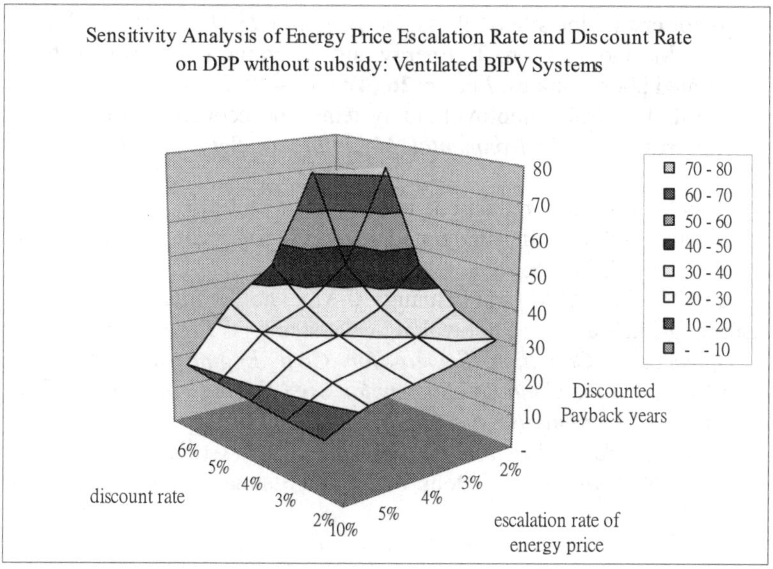

Figure 3: Sensitivity analysis of energy price escalation rate and discount rate on DPP without subsidy: Ventilated BIPV systems.

5 Conclusion

The benefits of BIPV system mainly derive from energy cost savings, enhanced power quality and reliability, reduced environmental emissions, rebates, and other incentives. The paper uses the empirical data and the economic evaluation model to explore the future of BIPV development in Taiwan. After examining critical factors after the PVs-project, the sensitivity analysis of the future PVs-project was shown. Based on the empirical residential case analysis located at Kao-Hsiung with a reasonable PR value, the results of sensibility analysis showed that BIPV will become one of the dominant renewable energies in the near future if integrating good building design is available. With higher escalation rate of electricity prices and lower discount rates, discounted payback years of BIPV systems will be less than 20 years. Adopting ventilated-BIPV will lower down the DPP threshold even more.

Acknowledgement

Support from the National Science Council of ROC through grant NSC 96-2221-E-426-008 in this study is gratefully acknowledged.

References

[1] Oliver, M. and Jackson, T. The market for solar photovoltaics. *Energy Policy* **27**, 371–385, 1999.

[2] Oliver, M. and Jackson, T. The evolution of economic and environmental costs for crystalline silicon PVs. *Energy Policy* **28** (14): 1011–1021, 2000.

[3] Oliver, M. and Jackson, T. Energy and economic evaluation of building-integrated photovoltaics. *Energy* **26** (4): 431–439, 2001.

[4] Lesourd, J-B. Solar photovoltaic systems: the economics of a renewable energy resource. *Environmental Modelling & Software* **16** (2): 47–156, 2001.

[5] Maycock, P. Cost reduction in PV manufacturing. Impact on grid-connected and building-integrated markets. *Solar Energy Materials and Solar Cells* **47**, 37–45, 1997.

[6] Hsu, K-J, Lai, C-M and Chiang, C-M. The Economic Evaluation of Building-integrated Photovoltaic Systems. *Proceedings, Annual Conference - Canadian Society for Civil Engineering, v 1, Annual Conference of the Canadian Society for Civil Engineering 2007: Where the Road Ends, Ingenuity Begins*, pp. 390–397, 2007.

[7] ASTM, *Standard Practice for Measuring Life-Cycle of Buildings and Building Systems*. Annual Book of ASTM Standards, Vol. 04.11, January 1995, America.

[8] Eiffert, P. *Guidelines for the Economic Evaluation of Building-Integrated Photovoltaic Power Systems.* Technical Report, NREL/TP-550-31977, National Renewable Energy Laboratory, 2003.

[9] Ruegg, R.T. and Marshael, H.E. *Building Economics: Theory and Practice.* New York: Van Nostrand Reinhold, 1990.

[10] Terpenny, J.P., Ambs, L.L., Dixon, J.R., Sullivan, J.L. and Sullivan, W.G. The Engineering Economics of Energy Use and Capital Investment. *Proceedings of the 2003 ASEE Annual Conference*, 2003.

[11] B. Marion, B., Adelstein, J., Boyle, K., Hayden, H., Hammond, B., Fletcher, T., Canada, B., Narang, D., Shugar, D., Wenger, H., Kimber, A., Mitchell, L., Rich, G and Townsend, T. Performance Parameters for Grid-Connected PV Systems. *31st IEEE Photovoltaics Specialists Conference and Exhibition*, 2005.

[12] The German Solar Energy Society (DGS), *Planning and Installing Photovoltaic Systems: A guide for installers, architects and engineers*, James & James Science Publishers Ltd. UK: London, 2005.

[13] International Energy Agency, *Photovoltaics in Buildings: A Design Handbook for Architects and Engineers*, James & James Science Publishers Ltd, 1998.

Ways to deal with the 'temporary' value of cost benefit analysis

J. M. Vleugel[1] & E. J. Bos[2]
[1]*OTB Research Institute, TU Delft, The Netherlands*
[2]*LEI, The Netherlands*

Abstract

A key problem with social cost benefit analysis (CBA) is that ex ante and ex post analyses of the same (infrastructure) project in many cases have different outcomes. Ex ante analyses are used to decide whether new infrastructure has sufficient net benefits for society, provided that a set of assumptions has been met. If this is the case, and sufficient budget is available, it is likely that a project to build this infrastructure is initiated. Ex post analyses are used to evaluate the real impact of the new infrastructure. This divergence between mentioned outcomes may lead to serious debates in politics and society, especially in cases where the ex post impact of a project is much lower than the ex ante calculations predicted. This is frequently the case for railway projects. In this paper we deal with this issue by discussing the main causes of this divergence and suggest some practical solutions. The main conclusion is that the way infrastructure projects are planned and developed may need reconsideration, looking at, in particular, environmental effects. This paper is based on ongoing studies.

Keywords: cost benefit analysis, environment, transport, decision-making.

1 Introduction

Cost-benefit analysis (CBA) "is a practical and rigorous means of identifying, targeting and checking the impacts of regulatory measures on the underlying causes of the ills with which regulators need to deal, those causes being the market failures that in turn may justify regulatory intervention" [1].

Intervention in this paper is restricted to public investments in infrastructure. Infrastructure is in most cases supplied as a public good, although it can also be supplied by private companies (in case of toll roads or privatized railway lines).

WIT Transactions on Ecology and the Environment, Vol 108,
www.witpress.com, ISSN 1743-3541 (on-line)
doi:10.2495/EEIA080171

CBA is a "tool to improve policy making, not an instrument to set policy goals or to determine which options decision makers should consider" [2]. The output of a CBA is the following: main problem, key assumptions, main option(s) considered, an appraisal to quantify the costs and benefits of these options, and if necessary involved trade-offs [3].

In the Netherlands, CBA is the main tool to support decision-making regarding public investments in infrastructure. In case of large investments, public bodies are by law obliged to commission a CBA. This means that the quality of CBA is of vital importance for decision-making regarding such projects. CBA is employed ex ante, which means that it is carried out before the construction of the infrastructure starts, hence also well in advance of the actual use of the infrastructure.

If a CBA would be carried out after the project is finished or when the infrastructure is already in use, this so-called ex post CBA may have a (significantly) different outcome.

In this paper, we will first elaborate a set of reasons why ex ante and ex post CBA may have different outcomes (see Section 2). This overview will then be applied to a real infrastructure case – the Betuwe dedicated freight railway line connecting Rotterdam port with its European hinterland (see Section 3).

Section 4 is devoted to the impact of project changes on investment cost, including those associated with the environment. Finally, Section 5 contains conclusions and recommendations.

2 The difference between ex ante and ex post CBA

2.1 Introduction

A CBA measures the welfare change due to the implementation of a project (plan scenario), in reference to the situation that the project would not have been implemented (autonomous development). In case of an ex ante CBA, the purpose of CBA is to support decision makers in making a choice between (variants of) a plan based on calculated impacts. Subsequently, effects have to be identified, quantified and expressed in economic terms. In case of an ex post CBA, the real impact of the (variant of a) plan that was realized is analyzed and only the autonomous development has to be assessed afterwards ('what would have happened if not').

Comparing ex ante and ex post is not common practice, given the different purposes of CBA in each case. However, it is common practice in politics and society to look back and question or even criticize past investment decisions in terms like '(in) adequate', '(un) sufficient', etc. This makes it in our opinion inevitable for science to elaborate this matter as well. Others are of the same opinion [4]: "Although we do not believe that the primary purpose of ex post CBA is to test ex ante estimates, we recognize that the results of ex post and ex ante CBA will inevitably be compared (if, as we assume, the results of ex post CBA are published). Such comparisons may have some value in helping to improve future ex ante analysis. It is therefore important to understand why differences between ex ante estimates and ex post figures might arise."

2.2 Towards an explanation of the differences: theory

Ex ante CBA is based on a set of assumptions regarding the future in which the foreseen infrastructure is being used. The amount of use of the infrastructure is one of the determinants of the long-term socio-economic impact of the infrastructure.

When developing infrastructure a key issue is to make a reasonable prediction about future demand for services using the new infrastructure. Many determinants play a role here, and the analyst has to make assumptions about each of them, and at least as important, about the interaction of these factors, because some factors are likely to stimulate demand, while others may work in the opposite direction.

Infrastructure has a considerable lifespan in which the factors influencing future use of the infrastructure may change significantly. For instance, traditionally main customers of freight railways were manufacturers of high mass goods. These goods, such as liquid bulk (e.g. fuels) and fixed bulk (e.g., steel pipes) are transported in high volumes. If an economy transitions towards a so-called post-fordian state, in which services become the dominant economic sector, then the demands on transport of manufactured goods may become much different. In such a society large-scale production is to a large extent replaced by smaller-scale tailor-made production. Finished goods are imported instead of being manufactured in the country or region. Goods are shipped in smaller batch sizes with a much higher delivery frequency. Stocking parts and goods is regarded as non-productive investment, hence reduced as much as possible. Society becomes dependent on reliable – just-in-time – transport services. Delivery by lorries is then the most practical transport solution, despite the fact that roads become increasingly congested and building new roads has become increasingly difficult, at least in the Western world. For other products, new opportunities for railways appear, like overland transport of (sea) containers with consumer products. Many of them are transported by road, but rail may be able to get a fair share, witness the frequent use of container trains.

Such transitions in the economy are an example of trends in the determining factors of transport, which cannot be easily be predicted for a longer period of time. All these development influence the demand for infrastructure and thus its benefits. Obviously, when practice differs from what was predicted, the ex post CBA outcome deviates from the ex ante CBA assessment.

Also concerning costs, ex ante assessments and ex post outcomes may diverge. Several factors may influence the uncertainty of ex ante assessment of investment cost. If a good is very specific the uncertainty of cost assessments increases. For example, a special type of bridge is likely to cost more than a standard type, but how much more is only known when it is finished. Experience with developing a product or service is another factor that has an impact on the (un) certainty of assessments. Lack of experience increases the uncertainty of assessments. This especially holds for assessment of the environmental impact of infrastructure.

Much of the biases concerning costs and benefits relate to the fact that information was not available on the time of the ex ante evaluation. However, it can also not be excluded that costs were mitigated in order to secure a favourite decision towards the project. Hiding information during ex ante assessment is known as client friendly consultancy.

3 The Betuwe freight railway line

3.1 Railways in transition

Rail freight transport has been declining for decades in Europe due to changes in energy supply (gas, oil and nuclear energy replacing coal), manufacturing and consumption. Financial deficits of the usually state-owned railway companies rose to an unbearable level and restructuring became inevitable. The role of the state with respect to (freight) railways changed. The European Union started a railway reform policy intended to create a competitive railway market. National governments split railway companies in infrastructure providers and service companies. Funding of the railway infrastructure remained a public activity in most countries, because private companies were unwilling to accept political risks (of changing or canceling railway projects) and had a relatively short time horizon with respect to the payback period of investments.

Two few major railway projects were initiated in the nineteen nineties in the Netherlands: the high-speed passenger railway line connecting Amsterdam, Rotterdam, Antwerp, Brussels and Paris, and the Betuwe dedicated rail freight line. The latter project will be discussed in this paper.

3.2 The Betuwe railway project

This project was initiated in a period in which several important issues emerged on the agenda of policy-makers in the Netherlands.

A first issue was the reinvention of the classical idea of international division of labour by authors like Michael Porter [5]. Following their ideas, every country should concentrate on specific competitive advantages. For centuries, The Netherlands is a major trading country. A specialization in transport and logistics follows from this. The national government translated Porters' ideas into a so-called mainport strategy, which concentrated on the development of the main transport networks and nodes, like harbours (Rotterdam, Amsterdam) and airports (Schiphol). Certain transport axes were regarded as main links to connect the countries main ports with the hinterland (Germany and other countries in Europe). Germany is the main importer of goods transported on Dutch roads, rail and inland waterways (by barges).

Among these main axes was a new, dedicated railway line for freight, connecting Rotterdam harbour with its German hinterland; the Betuwe rail freight line. In the same period (1980s–1990s), environmental and liveability issues were put on the agenda of politicians. The railway line enables rail freight operators to reroute a major share of rail traffic from the existing main east-west corridor, which crossed important urban areas, to this new railway line. The environmental and liveability impact of rail transport is regarded as lower, while traffic safety is higher than road transport. Then there was growing congestion on the rail network, leading to conflicts in scheduling passenger and freight trains. Growing congestion on the road network, and the peculiarities of barge transport (inland waterways do not cover the whole of Europe and use is depending on waterway conditions) did their share to stimulate the national government to develop this new railway line. The idea to develop this infrastructure came in 1989, while the infrastructure was put in operation in 2007 [6].

The project consists of 160 kms double track without level crossings. The line is electrified with 25 kV and equipped with the ERTMS safety system. To protect the environment and liveability the following was built: 160 fauna tunnels, 5 railway tunnels and bundling with the existing A15 main road. The infrastructure should last for some 100 years [7].

Regular traffic is growing, despite some technical issues related to signaling and safety of tunnels.

Figure 1 shows the line between the North Sea (Noordzee) and Germany (Duitsland).

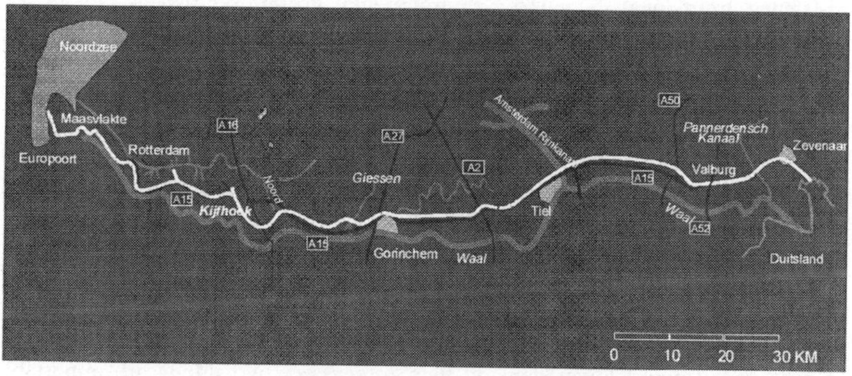

Figure 1: The Betuwe freight line. Source: [8].

Many studies have been carried out into the costs and benefits of this railway line, with results leading to positive and negative advice for the government. We will not discuss these studies in detail, but concentrate on some major uncertainties.

3.3 Uncertain benefits and costs

3.3.1 Benefits

A major uncertainty is related with forecasting demand. After decades of decline a trend change became visible in some countries in the 1990s. In the Netherlands the volume of international freight by rail has risen from 20 million tons in 1995 to 41 million tons in 2005 due to economic growth, internationalization of good production and consumption (globalization) and closure of German coalmines.

Second, institutional reform led to more between rail operators, offering more diversified, lower cost services [9]. Transport growth was concentrated mainly in containers, coal, other minerals, cars and steel.

The latest prognoses [9] for the year 2020 range from 34 to 91 million tons (or 75.000 to 140.000 freight trains) per year. This wide range indicates that the impact of the demand determining factors on demand may vary substantially over time.

Actual use affects the income of the rail operators and (via the user charges paid) also of the railway line owner.

3.3.2 Costs

Relevant costs can be divided into two main groups: pure infrastructure costs and costs made in order to protect the environment and liveability.

Uncertainties with respect to project costs may come in several ways. First, there may be technical changes to the project, for instance, the line may be extended; this is called a change of the scope of the project. Second, some disruptions may occur during building, like unknown or unstudied geological conditions, which may impact the routing of the project and building time. Third, resistance from local citizens or environmental groups (known as Not in my backyard or Nimby response) may force developers to change their plans. They have to invest in mitigating measures, such as building noise shields or tunnels or even rerouting (part of) the line in order to reduce the adverse impact on the environment. To some extent this is also a change of scope of the project.

Some of these changes and adaptations may be predictable, but not all of them. Their impact can even vary per case, which means that past projects, if at all similar are no direct basis for comparison.

Such changes are likely to increase the costs of an infrastructure project.

3.4 Reconsider or reject the project?

In case of a (much) lower demand than forecasted, two things are important. First, if the benefits are so low that they may not cover the costs of the project, a negative cost-benefit ratio will result. Second, the payback period of the investment becomes much longer or even infinite. Hence, both a social cost-benefit analysis as well as a private investment analysis would probably lead to at least a reconsideration or even rejection of the project.

Given the size of this paper, the paper concentrates on the costs of this project.

4 Impact of project changes on investment costs

4.1 Introduction

The impact of (1) technically and (2) socially induced changes on the Betuwe rail freight line was substantial. Table 1 shows what happened between the first cost estimates and the (nearly) finalization of the project.

Table 1: Development of costs for the Betuwe rail freight line (in mln. Euros).

Date/period	Extra	Total
1. June 1990: Cost estimate with 50% private financing		1.134
2. January 1992: Extension to Rotterdam-harbour (Maasvlakte)	+ 363	1.497
3. April 1992: Choice of trajectory Maasvlakte-Zevenaar	+ 838	2.335
4. May 1993: Adaptations to mitigate adverse effects on the environment (local resistance)	+ 497	2.832
5. September 1993: Mark-up for price inflation	+ 72	2.904
6. December 1993: Adaptation to parliamentary and provincial demands, Barendrecht now in the project	+ 335	3.239
7. September 1994: Mark-up for price inflation	+ 119	3.358
8. Budget Tracébesluit (trajectory decision) 1996; adaptations including 4 new tunnels	+ 386	3.744
9. Change of scope due to political decisions (e.g., Dintelhaven bridge, double-stack ready tunnels etc.)	+ 321	4.065
10. 1996-2005: Mark-up for price inflation	+ 783	4.848
11. Prognosis 1-1-2006 including risks	+ 3.714	4.653
12. Balance of higher and lower costs (temporarily)		–195

Note: All cost exclusive VAT.
Source: [10].

4.2 The differences between ex ante and ex post CBA

Following Table 1, four main sources of differences can be distinguished:

a) Change of the plan scenario (Table 1, points 2, 3, 8, 9): 1.908 mln. Euros.

This is a political decision.

b) Costs to mitigate the impact of the infrastructure on the environment. (Table 1, point 4): 497 mln. Euros. This is an (non consumer preference based) assessment of society's WTP to prevent adverse effects on nature.

c) Inflation (Table 1, point 5, 7, 10): 974 mln. Euros.

d) Project risk (Table 1, point 11): either budgeted, unspecified or 'hidden'.

The latest cost prognosis [11] mentions a sum about 4.700 mln. Euros, which actually implies a quadrupling of project costs.

What remains hidden in this official calculation is the impact of assessment bias, as discussed earlier. We may refer to an earlier paper [12].

4.3 Decision making

Political decision making is a complex process in which financial-economic, socio-economic and also political-strategic considerations play a key role.

Many major investment projects are carried out because politicians like to make statements. In order to safeguard a project, there is an inherent bias to overestimate the net benefits of a project and to underestimate its costs. If a project has a not so beneficial or even negative outcome, CBA is treated as one of the inputs of the process of decision making, but not the dominant one [13]. It may then be a task for economists to reduce this 'abuse' of CBA, e.g. by defining strict guidelines for CBA.

Once a plan to build new infrastructure is presented, and especially in case of so-called 'grand projects', like a railway line cutting through a country from east to west, it becomes apparent that not everyone is in favour of the project. In case of this project a vivid discussion started between national decision makers on one side and local and provincial stakeholders on the other side. Especially in the province of Gelderland there was a lot of resistance, because of the negative impact of the railway line on the landscape and liveability, the opinion that (their) inland waterways could accept the additional traffic without any problem and the (perceived) lack of local and regional-economic benefits.

The national government continued its support for the project nonetheless. In order to overcome this resistance major mitigation works were carried out (as mentioned in Table 1, point 4).

The cost of measures to mitigate adverse effects on the environment can be regarded as an indication of the social willingness to pay to prevent (more) damage to the environment. In case of the Betuwe rail freight line, this social willingness to pay equals almost half a billion Euro.

4.4 CBA and planning; some considerations

This project was not only confronted with strong opposition, but also a lobby started to stop it. But, the procedure to develop the project continued.

What, however, if the project would have been stopped at some moment? What would have been the economic impact of such a decision or the impact on liveability and the environment? Such questions have not been answered in a quantitative way. Nonetheless, they are very relevant. To use CBA in such a context may lead to adaptations in the way CBA is usually carried out. Table 2 gives examples of two CBAs carried out at different moments in time.

Table 2: Multi-step CBA.

	CBA_1 net value	Decision	CBA_2 net value	Decision
Project 1	< 0	No go		
Project 2	> 0	Go	< 0	Adapt or stop
Project 3	> 0	Go	> 0	Continue (limited adaptation)

In order to successfully apply such a step-wise framework, the project considered should have particular characteristics, more in particular, it should be changeable or split into separate stages, which are not linked in a technical sense (lack of indivisibilities).

5 Conclusions and recommendations

In this paper we compared ex ante and ex post CBA. Although such a comparison may be criticized given the different reasons for carrying out ex ante and ex post CBA, in practice it clearly makes sense.

The sources for deviations have been analyzed and quantified in economic terms for the Betuwe rail freight line case study. Four main sources are important: i) changes to the plan scenario, ii) (unforeseen) costs to mitigate the impact on the environment, iii) inflation and iv) project risk. From the referred case study it also follows that the willingness to pay for mitigating adverse environmental effects is 500 million Euros. This indicates that both Dutch policy makers and citizens consider the environmental impact of infrastructure a serious issue. As such they have a clear role in future CBAs for infrastructure in the Netherlands.

It may be a good idea to plan a project in a way that enables decision makers to later correct (presumably) 'wrong' decisions. CBA could then be carried out repeatedly and become a more dynamic tool to support decision-making.

Next to delivering financial benefits for society this approach may also be quite beneficial for the environment.

References

[1] Financial Services Authority (FAS), Practical Cost-Benefit Analysis for Financial Regulators Version 1.1, London, p. 5, 2000.
[2] Ibid, p. 1.
[3] Ibid, p. 19.

[4] NERA Economic Consulting, The FSA's Methodology for Cost-Benefit Analysis, London, pp. 26–27, 2004.

[5] Porter, M., The Competitive Advantage of Nations, Free Press, 1990.

[6] Ministerie van Verkeer en Waterstaat, Betuweroute, Voortgangsrapportage 22, Den Haag, 2007.

[7] See footnote 6.

[8] Keyrail, information provided, 2008.

[9] Francke, J., Oostroom, H. van, and Savelberg, F., Marktontwikkelingen in het goederenvervoer per spoor 1995–2020, Kennisinstituut voor Mobiliteitsbeleid, Den Haag, pp. 65–66, 2007.

[10] Prorail, De Betuweroute, slagader van het goederentransport per trein, Utrecht, 2007.

[11] See footnote 6.

[12] Vleugel, J.M. & Bos, E.J., Enige kanttekeningen bij kosten-baten analyse, paper presented at Colloquium Vervoersplanologisch Speurwerk 2007, 22 and 23 November 2007, Antwerp.

[13] TCI, Tijdelijke Commissie voor de Infrastructuur. Onderzoek naar infrastructuurprojecten. Reconstructie HSL-Zuid: de besluitvorming uitvergroot, Tweede Kamer, vergaderjaar 2004-2005, 29283, nr. 8, SDU, Den Haag, 2004.

Section 4
Natural resources management

An approach to a methodology for ecological valuation of environmental quality

J. A. P. Vásquez
Business Management School, EAFIT University, Colombia

Abstract

In Colombia territory planning depends on the natural resources on offer. Due to this offer and environmental quality, the most convenient uses of ground are fixed to satisfy the social demand, and the ecosystem sustainability. Often, natural resource quality valuation is managed from an analytical view, because it considers the natural environment as an independent sum of parts, that is water sources, ground, atmosphere, forest, amongst others. This way of natural resource management leaves apart the dependence relationship between natural resources, necessary for their sustainability. Ignorance of these relations implies serious implications in ground uses because relations between resources could be destroyed. This paper proposes an approach to a methodology for ecological valuation of environmental quality from an analytical and systemic point of view (considering relations between natural resources). The methodology allows us to estimate in a quantitative way the environmental values, such as intrinsic conservation value and relative conservation value, using environmental quality criteria and Colombia's environmental regulations.
Keywords: natural resources, conservation value, environmental quality, territory planning.

1 Introduction

Estevan [1] considers territorial planning as a basic element for Public Environmental Management, which allows environmental considerations in land use to be included in planning processes, then becoming a management instrument. Besides, it becomes an essential element to make decisions in relation to use, protection, defense, and improvement of natural resources.

Although Law 388, 1997 [2] has been a base for the country's territorial planning processes, it does not explicitly take into account a way to integrate the

WIT Transactions on Ecology and the Environment, Vol 108, © 2008 WIT Press
www.witpress.com, ISSN 1743-3541 (on-line)
doi:10.2495/EEIA080181

complexity of natural resources to as a system. Although Strategic Environmental Assessment allows a reflection on the environmental variable to be made in the territory planning processes developed by governmental entities, it does not explicitly propose a way to think about the environmental quality of natural resources of an environment from a systemic point of view.

For territorial planning from a perspective which intends to include environmental considerations, the existence of an environmental inventory (the natural resources on offer) becomes a basic element. As human interference effects on the territory depend upon an ecosystem's capacity to support man's activity, ecosystem vulnerability becomes a significant factor in a decision-making process concerning use, improvement, protection, and defense of the environment. Such vulnerability is also dependable on the conservation degree of natural resources in the ecosystem. The conservation degree of a natural resource will be a decisive factor for existence of all other resources of an ecosystem and environmental services which man can provide. Conservation degree determines then all functions a natural resource could offer to both the total ecosystem and man. If we consider natural resources conservation from a theoretical economic point of view, it could be thought that its conservation degree determines functions which the resource "lends" to both the ecosystem and man; for this reason, conservation has a very important social value. This article shows a methodology which integrates environmental norms and conservation values of natural resources to be applied when estimating environmental quality of an environment. The methodology's importance is a result of a combination made when integrating analytical and systemic thought and approaches to the environmental quality assessment.

2 Methodology

The proposed methodology is intended to estimate the conservation value of natural resources. For this purpose, this methodology is based on two basic elements: environmental legislation and an environmental quality unit system. The first element deals with legislation defined for a specific context of a country or region in which a territory's intervention programs or plans are included (in governmental plans). The second element refers to a system which allows taking different environmental indicators to the same measurement system, in such a way that they can be measured and analyzed in order to determine the conservation value of natural resources and to make decisions in relation to use, improvement, defense, and protection of resources.

3 Conservation value of a resource

This refers to the ecologic value which is given to a resource due to its particular characteristics or intrinsic qualities, as well as for "services" it "provides" to the ecosystem. This results in two conservation values: intrinsic conservation value and relative conservation value.

3.1 Intrinsic value

This deals with the existence value which is given to a resource due to its particular characteristics, such as conservation, deterioration, singularity, rareness.

3.2 Relative value

In a specific environment, this refers to a resource value according to its functions and services it provides for preservation of all other resources. Thus, water existence is vital for existence of other resources of the ecosystem, such as flora (forests and pastures), fauna, soil, etc.

In order to determine a resource's conservation value, quality parameters defined by environmental norms for each resource should be considered.

4 Environmental quality units

As each resource has different physical parameters which allow the determination of its conservation status, these parameters should be included in a unified system which makes it easy to compare quality parameters (indicators) of each natural resource with the conservation status of several resources of an ecosystem. For that purpose, transformation functions are used to express the status of a natural resource as environmental quality units.

5 Conservation value estimation

In order to determine intrinsic conservation value, a hydrospheric element will be depicted. This element includes all water forms which are present in the analysed environment, as well as its availability and quality. After indicators have been defined by environmental legislation, an environmental quality unit system is used to take contamination amounts into an E.Q.U. (Environmental Quality Unit) through some transformation functions proposed by Conesa [3]. From environmental quality units, the environmental conservation value of the place to be analyzed is determined. Table 1 shows a value estimation of water intrinsic conservation in three different basins which will be called A, B, and C.

When intrinsic conservation value has been determined for each natural resource of the territory to be analyzed, the importance of each natural resource should be determined (according to its intrinsic conservation value or its environmental quality) in relation to the existence of each other natural resource in such an environment. To carry out this procedure, quality indicators for each resource are taken. Table 2 shows relative conservation value of natural resources from C basin, in relation to all other resources of the environment.

6 Discussion on results

According to the results shown in Table 1, the basin having the best conservation status is basin B, which value (subtotal) in EQU and in relation to water resource

Table 1: Intrinsic conservation value.

Environmental Element	Measurement Unit	Basin		
		A	B	C
Hydrospheric				
OBD	mg/l	0,2	0,36	0,28
DO	%	0,98	0,28	0,2
SS	mg/l	0,35	0,8	0,14
Sub-Total (U.C.A.)		1,53	1,44	0,62

OBD : Oxygen Biochemical Demand, five (5) days.
DO : Dissolved Oxygen.
SS : Suspended Solids.

is 1.44, while basin C shows the smallest conservation value. In order to define which basin has the best conservation status, environmental quality units of all other natural resources in each basin should be determined, so the total value can be estimated by adding environmental quality units. Once the value has been determined, a comparison among the three basins can be made.

Intrinsic conservation value allows the determination of the vulnerability of an environment (for example, a basin) to an intervention plan, program or project, in such a way that human actions in such environment can be directed in order to protect natural resources, then becoming a possible support for environmental management.

In relation to the three basins, partial results would indicate that if human intervention is necessary in one of them, the best thing is doing it in basin C, given its lower environmental quality.

Now, from the concept of intrinsic conservation, environment is conceived as being formed by an isolated aggregate of natural resources (water, air, flora, etc.) without any interaction. Consequently, dependence relations are not recognized among natural resources which form an ecosystem. From this conception, the need for a concept which allows the consideration of dependence relations in an ecosystem arises. Relative conservation value intents to consider the interaction of natural resources in an ecosystem; for this reason, the relevancy of the concept is important for determining the ecologic value of the environmental quality of a resource.

In Table 2, the relative conservation value of a basin can be found in which it is possible to identify that the natural resource having the biggest relative importance is soil (geospheric component). Therefore, it is possible to conclude that if projects implying damage of soil resource (atmospheric element) are carried out in that basin - such as a road construction- harmful results for the ecosystem could be bigger than those for an intervention on other natural resources, for example in air (atmospheric element) or in water (hydrospheric component), etc.

Table 2: Relative conservation value.

Element Environmental	Atmosph	Geosph	Hydrosph	Biotic	Sum
Atmospheric					
Noise Level		1	1	7	9
Temperature		3	2	5	10
Rainfall		3	1	2	6
				Total	25
Geospheric					
Erosion	1		6	3	10
Phreatic Level	5		1	4	10
Slope	6		2	1	9
				Total	29
Hydrospheric					
OBD	1	1		8	10
DO	1	1		5	7
SS	1	1		4	6
				Total	23
Biotic					
Native	1	3	5		9
Birds (common)	1	1	2		4
Land	1	2	3		6
				Total	19

Atmosph : Atmospheric
Geosph : Geospheric
Hydrosp : Hydrospheric
Atmospheric: It refers to air quality and climate-related variables.
Geospheric: It refers to soil.
Hydrospheric: It refers to water.
Biotic: It refers to flora and fauna.

Consequently, relative conservation value is a supplement for the analysis resulting from intrinsic conservation value, because it allows guidance on basin C protection in relation to the type of human interventions.

For the aforesaid, it is possible to state that proposed methodology could become a support for territorial planning processes carried out by different governmental entities, as they allow the determination of the environmental vulnerability to human activities.

Likewise, it can be concluded that it is possible to achieve an approach to environmental quality issues of the environment from a systemic point of view through the use of quantitative methods which facilitate both analysis and decision-making processes in territorial planning programs, from an ecological vision.

As an approach, this proposed methodology is not intended to close the discussion on environmental quality assessment, but to encourage a debate on a very complex issue.

References

[1] Estevan Bolea, M. T. Master en Evaluación de impacto ambiental, La gestión ambiental, Volumen I, Editorial Instituto de Investigaciones Ecológicas, Málaga, pag 9–10, 1997

[2] Ministerio del Medio Ambiente, Ley 388 de 1997 Ordenamiento Territorial, Guía Legis, Colombia, 1997

[3] Conesa, Fernandez, V. *Guía Metodológica para la evaluación del Impacto ambiental*, Ediciones Mundi - Prensa, 3ª Edición, Madrid, pp. 253–300, 1988

Point to non-point phosphorus trading in the South Nation River watershed

D. O'Grady
General Manager, South Nation Conservation, Canada

Abstract

The South Nation River watershed has a regulated water quality trading program. By law, waste water dischargers must discharge 0 kg of phosphorus (P) loadings into receiving waters. New wastewater systems are now choosing trading instead of traditional P removal technology. These point source dischargers are buying P credits from rural landowners, primarily farmers. These credits are generated by constructing non-point source pollution control measures, and calculating the kg of P removed by each measure. South Nation Conservation, a community based watershed organization, is the broker for these P credits. The program is run by a multi-stakeholder committee, and all project field visits are done by farmers and not paid professionals.

Keywords: water quality trading, phosphorus trading, water credit trading, watershed trading.

1 Introduction

South Nation Conservation (SNC) is a community based watershed organization set up to manage the natural resources of the South Nation River watershed. Over the last several years, SNC paid over $1 million in grants to rural landowners for various non-point source pollution control projects.

The South Nation River watershed is located southeast of Ottawa, Ontario Canada. The 4,000 sq. km. watershed has a population of 125,000, and is mixed farming with dairy, cash crop corn and soybeans predominant.

The South Nation River has peak flows of over 1,000 cu. m./s. in the spring freshet, and less than 20 cu. m./s. during summer low flows. There are currently 16 wastewater lagoons in the watershed (14 municipal, 2 industrial). Provincial guidelines allow the lagoons to discharge their effluent at peak flows, primarily in the spring, for dilution of effluent to meet Provincial water quality guidelines.

WIT Transactions on Ecology and the Environment, Vol 108, © 2008 WIT Press
www.witpress.com, ISSN 1743-3541 (on-line)
doi:10.2495/EEIA080191

The South Nation watershed exceeds Provincial water quality guidelines for P, which is currently 0.03 mg/L. Annual mean P concentrations for the main South Nation River are 0.07 mg/L in the upper reach, 0.126 mg/L in the middle, and 0.129 in the lower reach of the River. Watershed studies show that 90% of the P load comes from non-point sources (NPS).

The Provincial Ministry of Environment (MOE) is responsible for water quality and licensing the operation of wastewater treatment in the Province. According to Provincial policy, where water quality does not meet Provincial standards for a specific contaminant, no further degradation of water quality will be allowed for that contaminant. However, in the past MOE gave dischargers a permit to discharge P from their plants into the South Nation River and its tributaries, even though the watercourses did not meet Provincial water quality objectives. Beginning in 1998, the Ministry stopped issuing these permits and required all dischargers to have zero discharge of P from their plants. MOE imposed this standard on new construction only. Existing plants that continued to operate according to their current permits required no changes to P loadings.

In the past, the only option for municipalities to meet this standard was improved wastewater treatment. However, it is not always technically feasible, physically possible, or socially desirable (because of costs) to meet the 0 kg standard.

MOE therefore allowed an innovative solution to remove P contributed by wastewater dischargers. Called Total Phosphorus Management (TPM), it allows dischargers to contribute P from their treatment plants, in contravention of Provincial policy, if they offset this increased P load by controlling P from non-point sources (NPS). To reduce P, the point sources buy P credits from non-point sources of pollution.

MOE treats the watershed as a unit. Since P is contributed throughout the watershed, it allows the TPM program to remove P anywhere in the watershed. A treatment plant discharging P in the lower reaches of the watershed can therefore pay to reduce P in the upper reaches, or any other part, of the watershed.

The number of kilograms of P to be bought depends on two factors. The first is the amount of P that the discharger contributes. For example, a village expanding their wastewater plant for an additional 3,500 people will add about 600 kg of P into the River. A recent landfill expansion added 25 kg of P.

The second factor is determined by the offset ratio. In theory, a discharger only needs to reduce P from non-point sources equivalent to the amount they contribute (i.e. a 1:1 ratio). However, MOE requires a 4:1 ratio for TPM. That is, 4 kg of P must be removed from non-point sources for every 1 kg of P contributed from a point source. This higher ratio is due to the unique nature of the TPM program (it is the first of its kind in Ontario), lack of knowledge on how much P is first transported, then delivered, to watercourses, and the debate on how much of the P in the water is soluble vs. particulate. The high offset ratio also allows a buffer in the event that a BMP is not 100% effective.

The amount of P contributed by various non-point sources is determined by formulae derived from studies in Canada and elsewhere. A study of the scientific literature published by South Nation Conservation in 2003 shows that the range in results for individual practices is quite large, and the results are highly variable since calculating P lost or saved by agricultural management practices is complicated. Since 2003 there have been other studies published on P reduction using various BMP measures, and these will be evaluated by SNC in the near future.

Following are examples of formulae used to calculate P removal from various NPS control methods.

1.1 Milkhouse washwater

Milkhouse washwater refers to the wastewater generated from cleaning the milking equipment, pipeline and bulk tank. It may include cleaning the milk parlour floor, which may contain manure, bedding and feed. P loadings depend on the number of cows, volume of washwater, type of milking system, detergents and management in the milking parlour.

P controlled by milkhouse washwater (excluding manure) projects
= # cows X 0.69 kg. P/cow/yr

P controlled by milkhouse washwater (including manure) projects
= # cows X 2.76 kg. P/cow/yr

1.2 Manure storage

The manure storage formula calculates P savings for proper manure storage. This may include construction of a concrete basin to replace stacked dairy manure piles, berms, a settling basin, or a buffer strip to treat feedlot manure. Two formulae are used: one for beef feedlots and one for dairy manure pile. It is assumed that P losses from feedlot manure are higher than piled manure since it is spread out and more exposed to rainfall, resulting in more clean water contamination.

P controlled by proper manure storage of beef feedlot manure
= # of animals X # days X P excreted X 0.30

P controlled by proper manure storage of dairy pile manure
= # of animals X days X P excreted X 0.07

1.3 Clean water diversion

Clean water diversions control manure runoff from barnyards, feedlots, and manure storage areas. It diverts clean water away from these areas using berms, eavestroughing or roofs and thus reduces P loadings in runoff. The number and type of livestock, size of yard and yard surface as well as the proximity to a watercourse will determine the amount of P delivered to a watercourse. The following calculation assumes that clean water diversion will control 50% of the P lost in runoff. The number of days refers to the number of days that manure or animals are on the yard.

P savings from clean water diversion for feedlot manure
= # animals X days X P excreted X 0.30 X (reduced feedlot runoff volume/original feedlot runoff volume)

for dairy pile, replace 0.30 with 0.07

1.4 Livestock access

Information on the input of P from direct livestock access to watercourses is rarely calculated. Studies generally group P non-point sources (erosion, manure) and do not break down P load reductions into its components. An estimated 3% of daily manure production is discharged directly into a stream when there is no alternative water source.

P savings from restricted livestock access
= # of animals X days X phosphorus excreted X 0.03

1.5 Septic systems

Only repair or replacement of biologically failed systems will produce a P savings that can be traded. Failed systems are often observed with ponding on the surface. P load savings for improved septic systems are:

P savings = P loading (failed) – P loading (functional),

Where P loading (failed or functional) = 0.6 kg. TP ca^{-1} $year^{-1}$ *X (# persons) X (1-A),*

Where A = attenuation in vadose zone (0 - failed; 0.4 functional sand; 0.7 – functional sand mixed with either silt, clay or red mud)

1.6 Conservation tillage

There is some debate in the literature as to whether conservation tillage increases or decreases P delivery from cropped fields. Overall it appears that conservation tillage reduces total P, although the soluble P delivery may increase. Studies show that conservation tillage reduces soil loss and total phosphorus. Various studies show that each hectare of cropland contributes 1 kg of P per year. A conservative 50% reduction is P is used for no-till.

P controlled per year by no-till = 0.5 kg X hectares
P controlled per year by cover cropping = 0.4 kg X hectares

1.7 Buffer strips

Buffer strips are areas of planted or naturally occurring vegetation that filters nutrients and sediments from agricultural runoff before it reaches surface waters. Buffer width is the most important factor in removing P.

P controlled per year by buffer strip = 0.67 kg X ha cropland buffered

2 Costing of P removal

The total cost of controlling each kg of P was determined by South Nation Conservation (SNC). Using the formulae above, it was possible to calculate the amount of P controlled for a number of projects recently completed by SNC. Since the total cost of each project was known, it was possible to derive an average cost of \$400 (Cdn) for each kg of P removed. Obviously, some projects are more efficient at controlling P than others, however the \$400 figure is accepted as accurate.

The \$400 also includes costs of project management (staffing, administration), water sampling, communications to promote the grants available to landowners, and yearly reporting. SNC must complete a yearly report showing the amount of P controlled that year, and allocating that P to each of the TPM dischargers.

Cost benefit studies done in the South Nation watershed show that the cost for complete removal of P using traditional wastewater treatment methods can be as high as \$20 million if a new treatment plant is required, to a low of several thousand dollars per kilogram if additional secondary treatment is added to a lagoon (sand filters, treatment wetlands, for example).

The advantages of using the TPM approach are evident:

1. It saves local tax dollars since new wastewater treatment plants are not required to control P.
2. It saves government dollars, since wastewater treatment costs are lower, and fewer government grants are needed.
3. It puts money in the hands of farmers.
4. It achieves greater water quality benefits since NPS controls will prevent not only P from entering the water, but other nutrients and pathogens as well.

Most TPM agreements have a four year deadline to reduce the full amount of P. Single municipalities have paid up to \$500,000 to SNC for P credits.

SNC is the broker between the point source and non-point source, handles all financial transactions between the two, and reports on compliance for P control. Publicity for projects is made through the local media and presentations to municipalities and farm organizations. All projects are voluntary, with no landowner forced to participate in the TPM program.

Neither SNC, nor the landowners as the recipient of the funds, have any legal responsibility should P targets not be met. This responsibility rests solely with the discharger who must prove to MOE that they meet their P reduction targets. With SNC's experience in grants for similar projects in the past, there is no forecasted shortage of P reduction projects for several more years.

The issue of responsibility for P reduction was a key issue to the success of the TPM program. Initially, the agricultural community opposed TPM. They had concerns with several components of the strategy including:

- low offset ratio for P reduction: initial ratio was 2:1
- low funding level per kilogram P removed: initial costs did not include sampling, reporting, communications, all administrative costs
- responsibility of landowners who accept funding to complete non-point projects: it wasn't clear if farmers were to be blamed

- responsibility of the municipality/industry if the P offset was not achieved through non-point source reduction projects.

The agricultural community felt dischargers had a license to pollute, and that the public would perceive that farmers were the cause of the problem if they were doing all the work and getting all the grants.

Extensive consultation with the agricultural community over three years achieved consensus on the roles and responsibilities for the various partners involved in the TPM program. This consensus became a *Statement of Roles and Responsibilities* document that was signed by the local agricultural organizations, Provincial government, and SNC. Consultation also resulted in a higher ratio for P reduction, higher costs per kg P removed, improved water quality monitoring and an overall program evaluation after 5 years.

While water quality trends show a reduction of P, it is not possible to attribute this solely to the TPM program. Watersheds are complex ecosystems, and water quality changes can be attributed to a number of factors, including fertilizer prices, rainfall changes, different tillage practices, etc.

However it is accepted that the P reduction targets are being achieved. The mathematical formulae are accepted as sound science since they originate from peer reviewed literature.

The consultation process also created a multi-stakeholder Clean Water Committee that approves all projects. The Committee is composed of farmers, industry, municipalities, farm organizations, and SNC. It reviews projects, and whether or not they meet the criteria for funding. All criteria, grant rates, and other water quality decisions are made by the committee. The Committee receives funding from several different sources, all with slightly different funding criteria. It then decides if the landowner project meets the criteria for one of the grant programs.

A final result of the agricultural consultation was the use of farmers as field representatives to do all site visits. The agricultural community expressed some concern over using agency staff who might not understand current farming practices. Now, when a landowner applies for a grant, they contact SNC, who then refers the call to one of several Farmer Field Representatives who then do the field inspection. These Representatives review the project and potential grants with the landowner, and determines if it is indeed eligible for grants. The Field Representative then makes a presentation to the Clean Water Committee, who rank projects based on improvements to water quality. This approach is rare amongst other water quality programs in North America, which tend to use full-time professional staff to do field inspections.

3 Conclusion

A point to non-point trading program exists in the South Nation Watershed. South Nation Conservation, a community-based watershed agency, acts as a broker for the program. Wastewater treatment plants are required, by regulation, to discharge 0 kg of phosphorus for new or expanded wastewater treatment plants. The amount of phosphorus removed is calculated using mathematical formulae, and dischargers must remove 4 kg of phosphorus for every 1 kg discharged into watercourses. The use of a trading program improves water quality over traditional wastewater treatment other pollutants are removed by the non-point pollution control methods, and not just phosphorus. It also reduces costs to all levels of government since a trading program is a low cost alternative to phosphorus removal. Finally, trading puts more money into the hands of farmers to improve the environment.

The agricultural community had initial reservations on the program, since they felt they would be blamed if phosphorus reduction targets were not met. However a signed agreement between farmers and regulators placed responsibility with the wastewater discharger.

The program is delivered by full-time farmers, who perform all interactions with other farmers. All decisions are made by a multi-stakeholder Clean Water Committee.

Marble quarrying: an energy and waste intensive activity in the production of building materials

V. Liguori[1], G. Rizzo[2] & M. Traverso[2]
[1]*Dipartimento di Ingegneria Strutturale e Geotecnica, Università degli Sudi di Palermo, Viale delle Scienze, Palermo Italy*
[2]*Dipartimento di Ricerche Energetiche ed Ambientali, Università degli Studi di Palermo, Viale delle Scienze, Palermo, Italy*

Abstract

Marble represents an important component of Italian buildings, where it is often utilized as a covering for bottom surfaces, despite its relatively high price. Moreover, it characterizes several public buildings, for which it is by far the most important decorative material, also because of its structural features and its long durability. Unfortunately, marble quarrying is an energy intensive activity that requires relevant amounts of electric and thermal energy sources; in addition, the extraction of the marble blocks from the mountain sides does involve a noticeable quantity of explosives, particularly in sites where traditional working methods are utilized. Another important feature of the marble mining is represented by the high level of waste materials released during the quarrying process. Both these elements call for careful attention to the production of this material, aiming for a suitable reduction of the environmental impact exerted by the current working procedures. An energy audit analysis, moreover, could allow the singling out of the steps of the whole process where it would be possible to reach improved efficiency, in this way properly cutting the energy resources involved in the production of the functional unit of this natural stone. The feasibility of such considerations is verified by means of an application to a marble quarry in Sicily, the region where an important rate of the Italian domestic production is realized. The field energy audit, other than suggesting a general approach to the problem, does indicate the high inefficiencies actually present in the working chain of the Sicilian marble.

Keywords: marble, life cycle assessment, energy and environmental audits, embodied energy, eco-indictors.

WIT Transactions on Ecology and the Environment, Vol 108, © 2008 WIT Press
www.witpress.com, ISSN 1743-3541 (on-line)
doi:10.2495/EEIA080201

1 Introduction

The building sector plays a meaningful role in energy consumption and environmental impacts. This sector is often defined the "40% sector" [1], since its accounts for about 40% of the whole energy consumption. For this reason it is important to adopt good policies and procedures for reducing the environmental impact of this particular sector. In this field, Life Cycle Assessment (LCA) established a strategic role for planning, monitoring and control of both energy and environment fallouts.

Marble production is one of the most important sectors in Italy and in particular in Sicily. In Italy its whole production accounts for 18% of world output; Carrara, a province of Tuscany, and Custonaci, a municipality of Trapani province (Sicily), plays the most meaningful role in marble tile production [2]. In fact, 190 quarries are located in Carrara and 54 in Custonaci.

In normal operative conditions, the main impacts can come from air emissions, the sludge and the large amount of scraps.

This work intends to present some results of a LCA application in marble cycle production and to make a comparison between two of the most important marble producers in the selected areas: Perlato di Sicilia and Bianco Carrara.

2 Marble and its production

Marble is a limestone or dolomite stone which is sufficiently close in texture to permit it being polished. Many other ornamental stones - such as serpentine, alabaster and even granite - are sometimes designated as marble, despite the term it should be invariably restricted to those crystalline and compact varieties of carbonate of lime (occasionally with carbonate of magnesia) which, when polished, are applicable to purposes of decoration [3].

Right now the definition is even more general and includes, under the concept of marble, all ornamental stones. The use of marble goes back several centuries. It is common to find several examples of marble in old churches, buildings and other important historical monuments.

Up till the XVI century, the techniques used for extraction of marble had been directly inherited by the Roman quarrymen of the first century, before Christ, and consisted of the careful use of the subtle cracks which divide the different layers of marble. All work was carried out by slaves where, by using metallic chisels and wooden wedges which were inflated by water, were then placed inside the natural cracks and easily managed to separate the marble blocks from the mountain. This was possible because there was a large amount of available labour and unskilled workers.

With the arrival of explosives, the excavation procedures changed drastically and the Apennine landscape went through a profound change and several "ravaneti" appeared everywhere. They were formed by large build-ups of debris, which were witness to the large waste of marble products due to the explosions.

Step-by-step the industrial activity relating to marble extraction and processing begun, with important factories established for the cutting and polishing of slabs. These productive units concentrated themselves at the bottom of valleys in order to benefit from the hydraulic energy generated by the rivers.

Since the end of the XIX century the helicoidally wire, a metallic cable able to dig out the stone, substituted almost completely explosives and caused another visible change in the landscape. The mountain stopped being destroyed, leaving behind piles of wreckage, and begun to be literally "cut", sculpted with precision, creating surreal landscapes made of huge flights of steps, and platforms called quarry warehouses where the stone is cut and prepared for transport.

Today the production is more technological and its life cycle mainly includes (figure 1) [4] the extraction in the quarry, the finishing treatment such as smoothing, polishing and finishing and the transportation and sale activities.

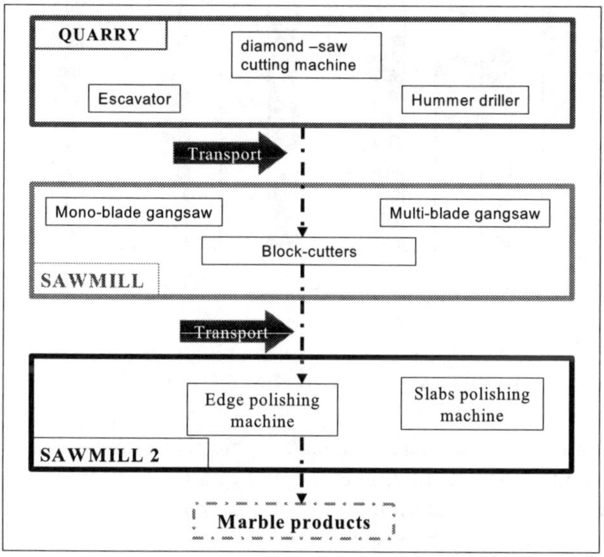

Figure 1: Flow chart of marble production.

3 Custonaci and Carrara marble basins

The extraction activity in Carrara goes back to Roman times. Indeed, the extraction of Custonaci marble started around 1950-51, mainly to cover local demand for white Carrara marble, which was becoming difficult and costly to supply. Anyway, in the Custonaci and Carrara marble fields, a wide amount of marble products are currently extracted (figure 2).

There are three important marble fields in Massa Carrara [5], the Torano basin, Fatiscritti basin and Colonnata basin.

The first one is a western marble basin, consisting of thirty-one quarries with a total production amount of about 27,000 tons/month. The quarried marbles are amongst the most valuable ones: Statuario, Statuario Venato and Calacata. The Fantiscritti basin consists of thirty quarries and produces more than 30,000 tons/month of marble products. The main products are: Bianco Ordinario and Venato. The Colonnata basin constitutes the eastern part of the Carrara marble region and holds about seventy quarries, 44 of which are still active, on a total surface of 500 hectares.

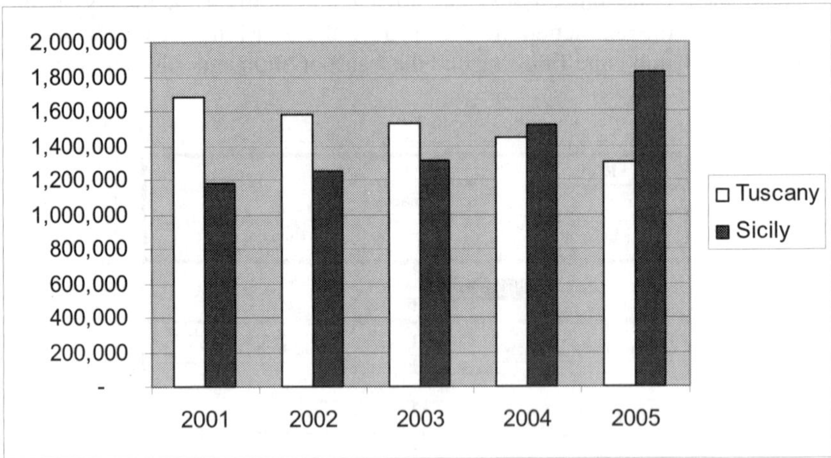

Figure 2: Comparison of marble production of Tuscany and Sicily.

The Carrara marble is made by metamorphic rocks consisting of 98% calcite and 2% dolomite, apatite, illite, goethite and quartz. They were originated from several phenomena of deterioration, compression, erosion, entrainment and deposit of detritus from pre-existent rocks and/or animal and vegetable remains in marine habitat.

The importance of this district is underlined by the current Italian standard that defines the Carrara Marble where it describes in details all characteristics that the marble has to have for being defined "*Carrara Marble*".

Also Sicily plays an important role in marble production in Italy. The various kinds of marble extracted in Sicily are shown in Table 1, with their main physics and geological characteristics.

Custonaci basin is consisting of about 54 marble quarries in a small area of 69 km^2. Due to its geological conformation, characterized by extensive outcrops of limestones of Mesozoic Dolomitic limestones, the territory of Custonaci is characterized by wide deposits of precious stone materials that have been exploited for fifty years and have contributed to the creation of a landscape marked by human presence where the dominant features are quarries. Presently, the Custonaci industrial area produces 1,800,000 tons/year that represents about 85% of marble in Sicily; it is equivalent to 15.7% in Italy and 2.7% in the world.

Only 25% of this material is used commercially, while the remaining 75% is constituted by waste of production.

Table 2 shows a comparison between characteristic data of two of the most important types of marble, that is Perlato di Sicilia (Custonaci basin) and Bianco Carrara (Carrara basin) [6].

Table 1: Kinds of marbles extracted in Sicily.

Name	Material kinds	Mining area	Manifacturing	Specific Weight (SW) [g/cm^3]	Resistance to compression (RC) [kg/cm^2]	Absorption [%]	Compressive Strength after gelivity [kg/cm^2]
Avorio Venato di Custonaci	limestone	Custonaci (TP)	smoothing and polishing	2.65	1,290	0.28	1,140
Pietra Lavica	basalt	Etna. Catania	smoothing and polishing	2.83	109	0.16	108
Granitello	limestone	Piana	polishing	2.68	1,230	0.26	1,215
Perlato di Sicilia	limestone	Custonaci (TP)	smoothing. polishing and bushammering finish	2.68	1,250	0.32	1,045
Pietra Sabucina	Calcarenite	Sabucina (CL)	Columns and capitals	1.33	130	7.5	92
Rosso Bolognetta	limestone	Bolognetta (PA)	smoothing. polishing and bushammering finish	2.67	1,215	0.36	1,087
Botticino Venato	limestone	Custoanci (TP)	smoothing				
Grigio Mirto	limestone	San Marco D'Alunzio (ME)	smoothing. polishing and bushammering finish	2.69	1,085	0.12	925
Pietra Pece di Ragusa	Asphalt Stone	Ragusa	smoothing and polishing	2.18	133	2.74	128
Rosso San Marco - Rosso antico di Sicilia	limestone	San Marco D'Alunzio (ME)	smoothing. polishing and bushammering finish	2.67	1,335	0.16	1,268
Grigio San Marco	limestone	San Marco D'Alunzio (ME)	smoothing. polishing and bushammering finish	2.68	1,220	0.11	1,036
Arenaria grigia dei Nebrodi o Quarzarenite	Quarzarenite	Nebrodi	smoothing and polishing	2.6	1,040	1.82	887
Grigio San Marco	limestone	Messina	smoothing. polishing and bushammering finish	2.68	1,220	0.11	1,036
Pietra di Cosimo		Comiso	smoothing and polishing	2.81	1,320	0.93	1,060
Perlatino di Sicilia	limestone	Custonaci (TP)	smoothing. polishing and bushammering finish	2.67	1,280	0.31	1,070
Grigio di Billiemi	Calcareous breccia	Palermo	smoothing. polishing and bushammering finish	2.7	1,430	0.12	1,226
Label Cream	limestone	Bolognetta (PA)	smoothing. polishing and bushammering finish	2.61	1,280	0.34	1,083
Nerello di Custonaci	limestone	Custonaci (TP)	smoothing. polishing and bushammering finish	2.61	1,088	0.21	984
Botticino di Sicilia	limestone	Custonaci (TP)	smoothing. polishing and bushammering finish	2.66	1,180	0.34	1,040

Table 2: Comparison of Custonaci and Carrara Marbles.

Name	Description	Specific weight (SW) [g(cm^3)]	Resistence to compression (RC) [kg/cm^2]	Compression strength after gelivity [kg/cm^2]	Water absorption coeff. [%]	Impact resistance [cm]	Frictional wear coefficient
Perlato di Sicilia	Cretacious calcareous, it's light ivory in colour with brown arabesques with darker or lighter shades and beautiful pure calcite streaks that recall the mother of pearl inside of shells. Suitable for any indoor or outdoor application in modern building and also in urban furnishing.	2.68	1,250	1,045	0.32	29	0.57
Bianco Carrara	Bianco Carrara "D" is a marble quarried in Carrara's area. It has grey background and dark grey veins. The quality is determined by the absence of large grey-black veins and quartzite concentrations. Suitable for any indoor and outdoor application and also for urban furnishing.	2.71	1,334	1,300	0.12	56	0.58

4 Energy and environmental audit of Custonaci marble

As has been previously pointed out, marble production has a very important landscape and environmental impact, particularly during the extraction and cutting steps. Where the marble quarry has a really large dimensions and usually where the geology characteristics of land allows to extract marble, there are several quarries in a small area. This can cause several environment problems, such as: disposal of scraps and sludge, along with air pollutant emissions. Generally speaking, during the extraction phase, the remarkable production of the so called "ravaneti", that is the scraps produced during the quarry operations, represent by far the most important wastes. On the other hand, the cutting and polishing phases produce a large amount of slush, that are essentially constituted by a mix of marble dusts with the cooling water utilized in the working process; the solid part of this mix is called "marmettola" that, sometimes, can be usefully addressed toward the employment in the building and civil sector.

The energy audit carried out in this work is particular important because it involves one of the most productive and impacting activity sector in Sicily.

In a context of energy saving and reduction of CO_2 emissions, the proper understanding of the amount of these impacts does represent a strategic move for reaching the fixed environmental targets of the European Union.

The adopted impact assessment methodology is the so-called "problem-oriented" [7] in which, as stated by the ISO series 14040 [8], the inventory data are associated with specific environmental impact categories, in order to better understand those impacts. The result of the energy audit is shown in figure 3.

The contribution of the energy utilized for explosive purposes is very small in comparison with the other compounds, but it has to be suitably considered because of its environmental impacts.

The most relevant energy impact is produced by sawmills, where the marble is cut and treated to be transformed into slabs and tiles.

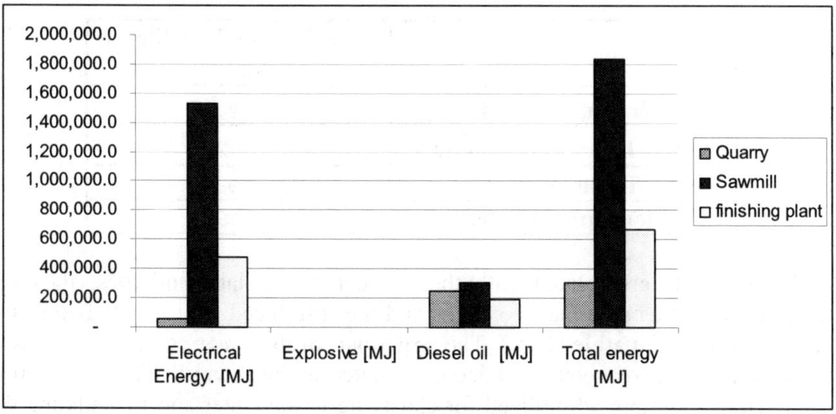

Figure 3: Energy inputs of marble production in the examined site.

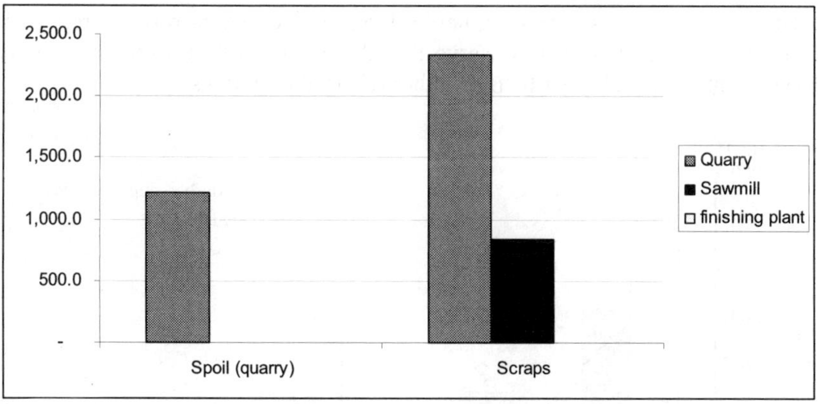

Figure 4: Main wastes from marble production in the examined site.

The result of total produced scraps is shown in figure 4. The quarry scraps, the so-called "spoil", are obtained using the explosives for moving the cut blocks from the mountain. This operation is necessary for allowing workers to transform the almost shapeless mass of stone into squared or semi-squared blocks, with standard dimensions. In fact, it is not possible to control the explosion very well that is used to move the big blocks from the rest of the mountain (since it depends on the characteristics of the deposit); so it often causes a big amount of spoil which is transferred directly toward a dedicated landfill.

Other inputs, such as water and resins, along with their amounts are reported in table 3.

Table 3: Other material inputs of marble production.

	Marble-chip floor tiles & slabs
Water in quarry [m³/m³]	0.21
Water in sawmill 1 [m³/m³]	0.12
Resin [Kg/m³]	3.99
Floculating [Kg/m³]	0.19

These inputs are related with the production of slabs and tiles: they are referred to by the cubic meter of slabs and tiles produced, that is assumed as the *functional unit* of marble (1 m^3). The larger part of the water comes from private draw water and is recycled by a depurator, sited in the sawmill. The flocculating is a chemical compound, utilized for clarifying waste water and flocculating the sludge. Then the sludge is filtered by a filter-press.

The resin is used to improve the resistance of slabs and tiles before the finishing treatment takes place.

Greenhouses gasses produced, here referred to CO_2 emissions, are reported in figure 5. It is interesting to observe that the amount of CO_2 emissions is two orders of magnitude bigger than the other released pollutants.

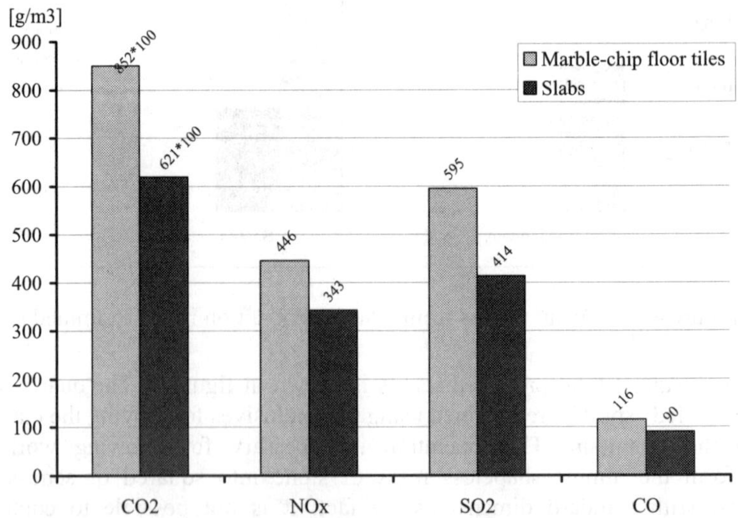

Figure 5: Pollutant emissions from the marble production in the site.

5 Results

The environmental impacts which have been taken into account in this study are energy consumption, water effluents, soil and air emissions and resources consumption. All results are reported in terms of functional unit (m^3) of products in order to allow a comparison with similar international and national data. Main products in the plant analyzed are slabs and cheap-floor tiles.

The index here adopted to compare the result of energy consumption with other building products is the embodied energy (EE) [9]. It is defined as the energy consumed by all of processes associated with the production of a building material, from the acquisition of natural resources to product delivery. The results of this work are reported in figure 6. It is evident that marble tiles are more expensive in terms of energy than slabs, while the biggest share of energy is attributable to the electrical energy. The energy for explosive is very small in comparison with the other two compounds but its use produces a remarkable amount of soil spoil.

The last part of the environmental audit procedure refers to the effects of the environmental emissions into some relevant damage categories [10], that allow the evaluation of the environmental impacts by three different points of view:

- Human Health
- Ecosystem Quality
- Resources

The results for the Perlato di Sicilia are shown in the figure 7: the biggest impact of functional unit of marble of Custonaci [m^3] is on the Ecosystem Quality category. This means that the impact is not directly related to the human beings but it can seriously compromise the natural ecosystems and can contribute to the climate change.

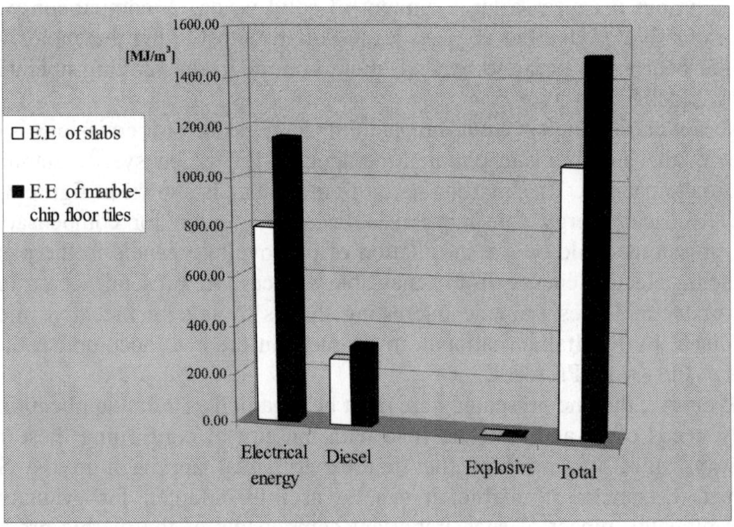

Figure 6: Embodied energy (EE) of the marble production in the site.

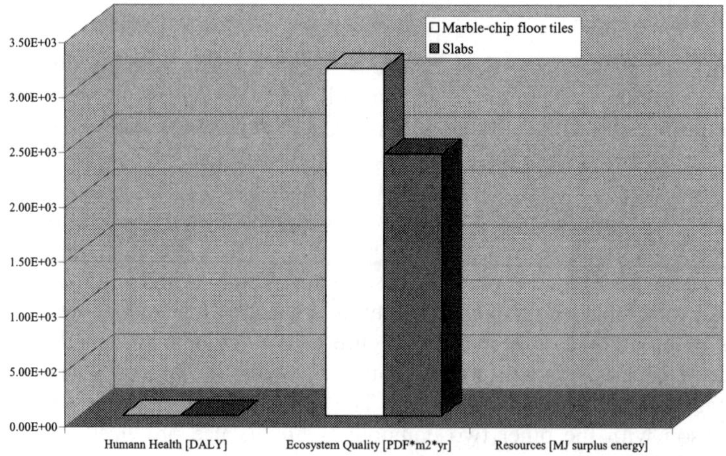

Figure 7: Eco indicator results of marble tiles and slabs.

6 Conclusions

The study presented in this work has reported the energy and environmental audits of marble slabs and tiles produced in the Custonaci basin, along with a simple comparison with that produced in Carrara. In fact, it represents an important national problem in Italy which is why it is really important in singling out the weak points of marble's production cycle in terms of energy and environmental efficiency.

The results indicate that the use of explosives is not so meaningful for air emissions but it causes a large amount of solid waste. A comparison with the Carrara marble production suggests a solution, by substituting the explosive with straddle bearing in order to achieve more control in the moving operations of marble blocks.

Moreover, the biggest environmental impact regarding air emissions is carbon dioxide, that is a relevant pollutant component for the ecosystem stability and for climate change. The biggest share of emissions, as shown in figure 3, come from electrical energy consumption, so a good solution for compensating the CO_2 emissions could be the installation of photovoltaic panels in the plants for producing electric energy from renewable sources. At present, access to these kind of technologies is made interesting thanks to several incentive measures introduced by the Italian national government in order to accomplish the main goals of the Kyoto Protocol.

Of course, the one presented here must be regarded as a first application and it surely needs other audits in the field with the aim of confirming these results. Anyway, since it represents the first experimental approach in the Sicilian productive context of marble, it can be usefully adopted for evaluating the environmental impact of such important components of the marble production, that is the "Perlato di Sicilia".

References

[1] CAN Europe. Input from environmental NGOs at the start of next round of the European Climate Change Program (ECCP), Climate Action Network Brussels, 24th October 2005.

[2] Nicoletti G.M., Notarnicola B. and Tassielli G. Comparative Life Cycle Assessment of flooring materials: ceramic versus marble tiles, Journal of Cleaner Production N. 10 2002, pp. 283–296, Elsevier, 2002.

[3] UNI EN 12670:2003, Natural Stone: terminology.

[4] La Gennusa M., Raimondi C., Rizzo G., Traverso M. Environmental impact of marble mining: the case study of a Sicilian marble quarry. Proceedings of SETAC Europe - 13th LCA Case Studies Symposium. Stuttgart, Germany, 7-8 December 2006.

[5] Bradley F. Le Cave di Marmo di Carrara. Guida ai materiali e produttori – Studio Marmo, Imm Carrara, pp. 110, 1998.

[6] Pieri M. Marmologia. Dizionario di marmi, graniti Italiani ed Esteri, Hoepli editore, Milano, 1966.

[7] Heijungs R. Environmental life cycle assessment of products: guide and backgrounds, Leiden University, the Netherlands. Center for Environmental Science; (CLM), Leiden University, the Netherlands, 1992.

[8] ISO, International Organization for Standardization. ISO/DIS 14040: Environmental management — Life Cycle Assessment — Principles and framework. 1997.

[9] Lenzen M. and Treloar G.J. Embodied energy in buildings: wood versus concrete, reply to Börjesson and Gustavsson, Energy Policy, Vol 30, pp. 249–244, 2002.

[10] AAVV,. The Eco-Indicator 99: A damage oriented method for Life Cycle Impact Assessment. PRè Consultants B.V., 2000.

Sustainable management of artisanal fisheries in developing countries; the need for expert systems: the case of the Pêchakour Expert System (PES)

S.-C. Chakour
Jijel University, Algeria

Abstract

Environmental problems are very often ascribed to pollution. Nevertheless, various other environmental impacts are also associated with abusive exploitation of natural resources that remain abundant in many developing countries. An urgent public intervention Boncoeur [1], therefore, seems necessary and reasonable public policies must be found. Since economic resources and coastal environment in some countries, notably in developing ones, lack expertise, the recourse to the Expert System triggers public attention and intervention for sustainable management of resources. In this respect, the present article examines advantages of expert systems, notably the Pêchakour Expert System (PES), as being helpful tools in decision-making. Another objective of this article is to highlight the efficiency and practical character of such expert systems.

Keywords: marine resources, artisanal fisheries, sustainable management, public policies, decision, Pêchakour Expert System.

1 Introduction

The development of the fishing sector in Algeria calls for an appropriate adjustment of fisheries in the context of sustainable management of resources and fisheries activities Chakour and Allegret [2]. This must rely on good governance based on indicators and relevant information, so as to orientate the public intervention. As a precaution, it is also necessary to use simulations that can put in evidence the after-effects of all economic or technical activities.

WIT Transactions on Ecology and the Environment, Vol 108, © 2008 WIT Press
www.witpress.com, ISSN 1743-3541 (on-line)
doi:10.2495/EEIA080211

Considering the difficulties facing managers of the fishing sector, either on the methodological plan or the conceptual one Le Gallic [3], the use of expert systems can undoubtedly contribute to resolving the problematic issue of the sustainable management of the sector in question, notably in developing countries. In this end, the present article studies the advantages of expert systems, notably the Pêchakour Expert System (PES) as being helpful tools in decision- making. Another objective of this article is to underline the efficiency and practical character of such expert systems.

2 Presentation of the PES

It is necessary to recall that the expert system is, conceptually, founded on the bioeconomical model "Pêchakour" Chakour and Boncoeur [4] which is theoretically justified on the basis of economic resources.

2.1 Organization and functioning of the PES

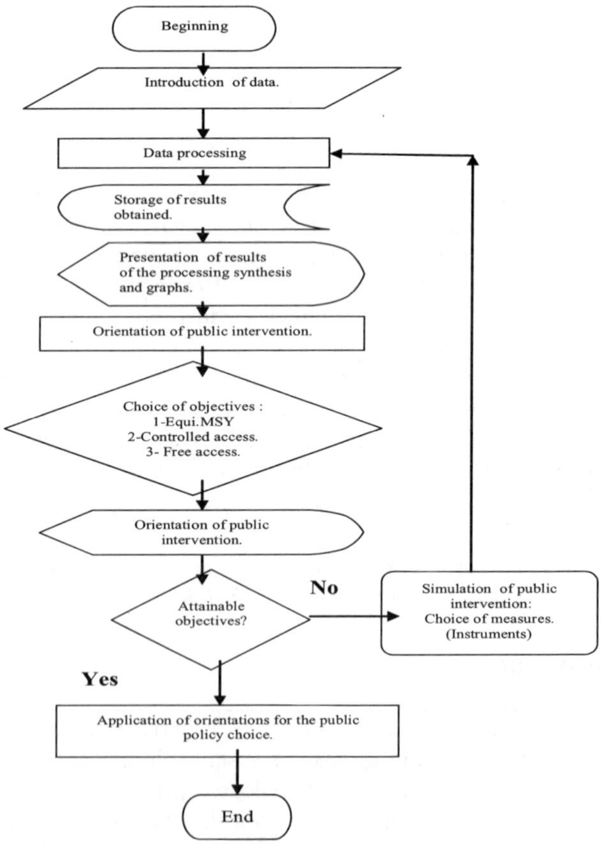

Figure 1: Summary of the functioning of the PES.

3 Contribution of the PES to the management of artisanal fisheries in Algeria

Methodologically, the PES is about replicating the real conditions of fishing activities, in a fishery reserved to the "small professions" in the gulf of Ziama. While leaning on results of an empirical approach that has permitted the identification of some technical and bioeconomical indicators, the PES Chakour [5] will be used for the management of the fishery in question.

Table 1: Necessary data introduced.

Type of information	Data	Definition	Value	Unit.
Biological	*a*	Biological parameter of the captures function.	0.0002	Coefficient
	b	Biological parameter of the captures function.	1103	Coefficient
Economical	*Im*	Middle value of acquisition of a fishing unit.	3000	Thousands of dinars
	Taxes	It is the total value of taxes on the whole of the flotilla and the marine strength by period of analysis (month).	09	Thousands of dinars by period (month)
	Insurance	It is the total value of insurances on the whole of the flotilla and the marine strength by period of analysis (month).	05	Thousands of dinars by period (month)
	Role	It is the total value of roles for the whole of the flotilla and the marine strength by period of analysis (month).	08	Thousands of dinars by period (month)
	pe :	Cost by unit of effort.	03	Thousands of dinars by unit of effort (exit)
	pc :	Price of sale of a captured unit.	300	Thousands of dinars by ton
	a :	Expresses the part of endowments independently to amortizations of the effort.	0.001	Rate between 0 and 1 : [0, 1].
Technical	*N*	Size of the registered flotilla exercising in the fishery.	40	Number of fishing unit.
	t	The length of life, in year, of a fishing unit (Boat)	20	Years.
	S	The average number of Units of effort (output) by unit of fishing (vessel) and per year. [1]	150	Exits per year.
	Period of analysis.	The period of analysis is tributary of the period of observation in the case of empirical approaches.	month	The month.

[1] ***S*** and ***t*** serve to calculate ***n***: length of life in equivalent effort. It is about the number of effort units, necessary to amortize a unit of fishing (vessel). With: $n = S.\ t$.

3.1 Source of data

The data below has been gathered from the empirical approach Chakour [6] where an investigation and the follow-up fishing activities are exercised by the flotilla "*small profession*" (the term "Small profession" for the artisanal fishing activity exercised by the small crafts or the small vessels will be use throughout this presentation. It is the integral translation of "Petits métiers" in French) in the gulf of Ziama.

The introduction of technical, biological and economical data is the most important phase. The data will be will validated and processed .The storage of information will be displayed on the user's demand.

3.2 Results of PES Processing

3.2.1 The synthesis

The *PES* allows for the displaying of three types of curves, including a curve of synthesis. These allow, in turn, the presentation of three functions: the total income function (RT), the total costs function (CT) and the total profits function (Pi). Information on the different balances is tabulated. The following two figures summarize the results of the processing.

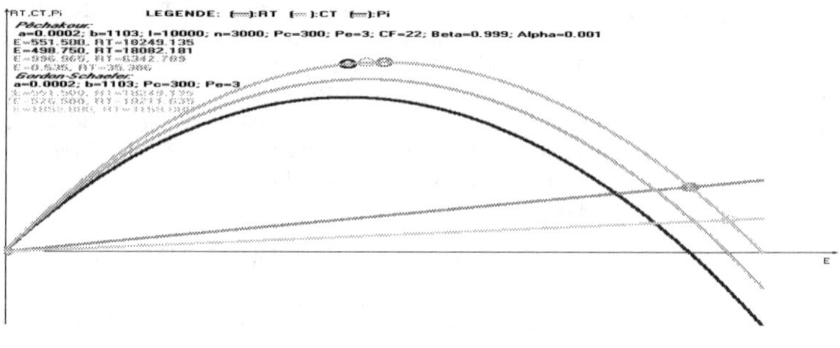

Figure 2: Representing curves.

The diagram and the summary table above are important references for managers in that they not only inform us on the different balances, while proposing a comparison between the Gordon–Schaefer model Gordon [7] and Schaefer [8] and the Pêchakour model, but also show levels of incomes and profits corresponding to the different balances. However, this information is not enough in itself, and a second processing consolidated by susceptible interpretations to orientate public intervention is necessary. Such a task is now possible because of the so-called *PES*.

3.2.2 Contribution of the *PES* to the orientation of public intervention

The orientations presented below come from the processing of information.

In terms of public intervention, the *PES* offers the choice between:

a. Action on the limitation of the fishing effort by endowing the limits not to exceed, the latter will serve for the delimitation of the size of the flotilla “small professions” being exercised in the Gulf of Ziama,

b. Action on the system of quotas by giving the levels of desirable captures “admitted” that are related to every balance and the quotas by the corresponding fishing unit.

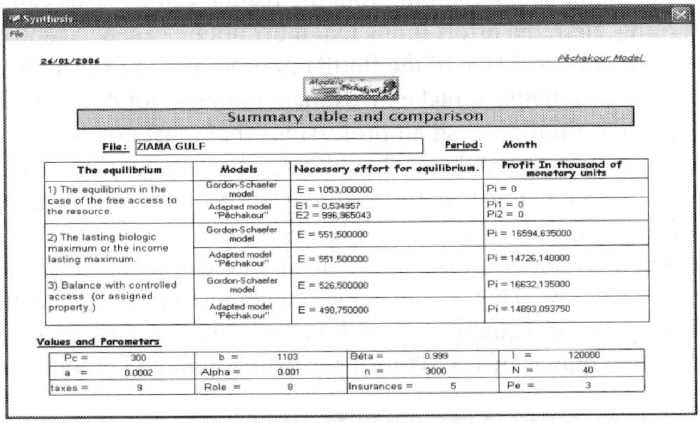

Figure 3: Summary of results.

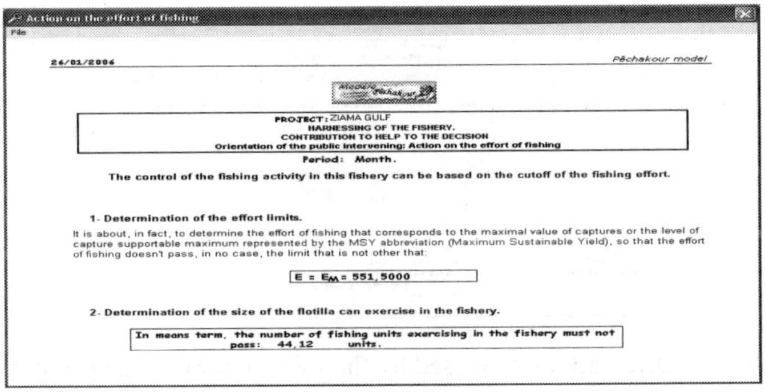

Figure 4: Orientations proposed by the PES: action on the effort of fishing.

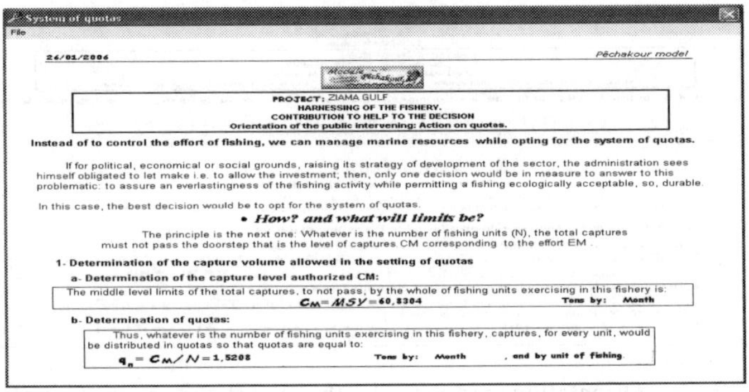

Figure 5: Orientations proposed by the PES: system of quotas.

3.2.2.1 Orientations within the sustainable biological maximum

- Within the balance linked to the sustainable biological maximum Clark [9], the rhythm of exploitation must be respected by avoiding to bypass "the limit effort" which must not exceed E = 551.5 units of effort per fishing unit and month. In fact, managers can delimitate the size of the flotilla which must not bypass, in this case, a total of 44.12 fishing units. Therefore, the main orientation is based on limiting the fishing effort. This can be done by determining, first, the effort limits that must not be exceeded and, second, by determining the ideal size of the flotilla in order not to exceed the MSY. The information also helps to make investment policies, for example, in our case, the size of the flotilla i.e. small professions that must not exceed 44.12 units (44 units). Decision-makers should take the amount of this limit into consideration knowing that by rejecting all projects of "small profession" investment in the site in question if limits were not be respected.
- If the action on the effort proves difficult, one can focus his decision on the system of quotas. This is done by recommending a delimitation of the monthly level of captures by a limit running to CM = 60.8305 tons, distributed in quotas of 1.5208 tons per fishing unit and month.

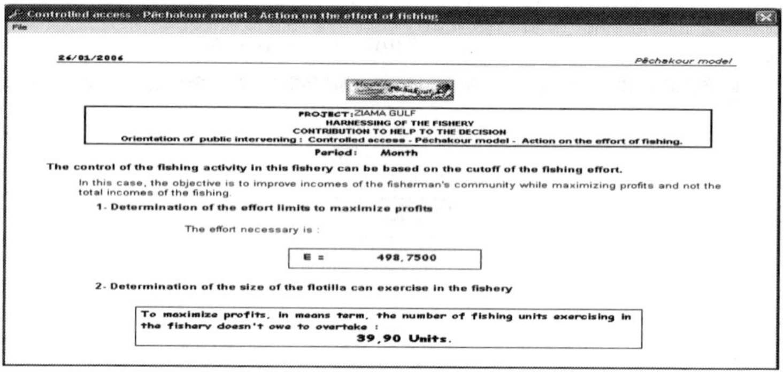

Figure 6: Orientations proposed by the PES: action on the effort of fishing.

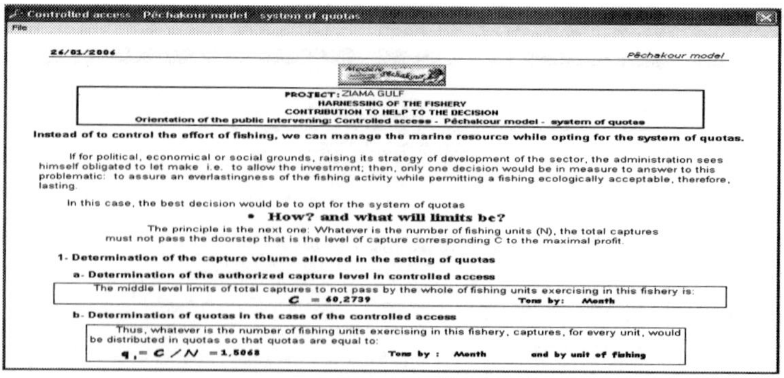

Figure 7: Orientations proposed by the expert system: system of quotas.

3.2.2.2 Orientations in case of controlled access to the resource

Orientations, within the controlled access, come from the hypothesis of profit maximization Gordon [10] while endowing the decision-makers of pertinent information on the profitability of the fishery. Indeed, the target would be the profits function. In the case studied, the necessary effort to maximize profits of the present fishery is 498.75 units of effort per month. Considering the equivalent of the effort exercised by a flotilla of a size of 39.90 fishing units i.e. "Small profession" excess of this limit, would lead to a reduction of profits. One can also resort to the system of quotas in order to maintain profits at a maximal level Mesnil [11]. To do so, he has to maintain the level of captures at about 60.279 tons per month. This can be achieved by opting for quotas running to 1.5068 tons per fishing unit per month.

3.3 Findings analysis

With reference to the obtained results, the ***PES*** shows a remarkable ability in treating and presenting both applicable and reliable information, as compared to those from the manual processing. In addition to this, the strength and the "advisor" character of the ***PES*** go beyond the mere presentation of information as it helps decision-makers taking appropriate and reasonable actions. Subsequently, this achievement is due to PES guidelines emanating from various situations and objectives.

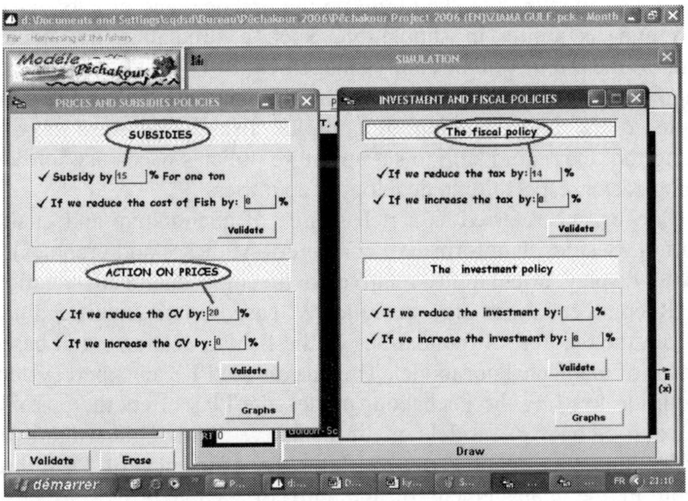

Figure 8: Action on the fiscal and prices policies.

3.4 Simulation of public intervention with the intention of developing policies

In addition to the purpose of public policies, the simulation of public intervention permits the consideration of scenarios through fixing objectives and evaluating impacts. In what follows, attention should be focused on showing the importance and functioning of the simulation operation. This can be done by fixing objectives and choosing the appropriate instruments.

3.4.1 Teachings withdrawn

As far as simulation is concerned, it is, therefore possible to combine the fiscal policies and investment along with the prices and subsidies policies. In this case, the induced effects would affect the whole system. In our case, action on subsidies and total costs remains one of the best combinations for the purpose of maximizing profits. Indeed, maximizing profits means the maximization of the gap between incomes and the total costs. Thus, all measures aiming at the reduction of costs, accompanied with an increase of incomes, would be at the origin of the improvement of profit levels. This justifies the recorded change as regards the balance, either in the case of free access or in the case of controlled access as well. Therefore, each increase in basic sale prices of fish or/and each decrease in total costs would lead to a new balance, either in the case of controlled or free access while demanding a more important effort, within the limits of the MSY in order to maximize profits. However, this may raise problems in the case of free access.

4 Conclusion

This study has attempted to shed light on the practical and efficient character of the *PES* and its ability in simulation. Such a simulation has become very important to the public policies and management of fisheries as well. In addition to pedagogical contribution, the use of PES in managing the Gulf of Ziama, will contribute to the understanding of complex fishing systems. It is, also, an interesting tool that could help to set up future policies necessary for developing the fishing sector, especially in developing countries.

The *PES* can be referred as a rich source of quantitative and qualitative of information. Besides its informative character (syntheses and graphics), the PES permits to display orientations relative to the bioeconomical situation of the fishery. Recommendations are, theoretically, based on an expert evaluation that draws from the theoretical foundations of the fishing economy and based on the hypotheses of the Pêchakour model. This makes the PES an expert system.

In addition to using the Pêchakour model, the PES offers the possible use of the Gordon–Schaefer model as informative and educational as well. Altogether, this test which aims to show the efficiency of PES through its utilization in the management of the gulf of Ziama proves to be suitable, appropriate and efficient at the same time.

References

[1] Boncoeur, B., Le mécanisme de la surexploitation des ressources halieutiques, Académie des Sciences rst n°17, décembre 2003, Ed. TEC & DOC, Lavoisier, 2003.

[2] Chakour, S.-C. & Alegret Tegero, J., Evolucion institucional y desarrolo del sector pesquero en Argelia; Palamos : Ayntamiento de Palamos, 2006, Collectccion : Quaderns Blaus, Espagna.18.Dec.2006.

[3] Le Gallic, B., Modélisation bioéconomique et gestion durable d'un système complexe de ressources communes renouvelables. Application au cas des pêcheries de la Manche. hèse de Doctorat de l'Université de Bretagne Occidentale, Mention Sciences Economiques, mai 2001.

[4] Chakour, S.-C. & Boncoeur, J., Un modèle bioéconomique pour une gestion durable des pêcheries en Algérie: le modèle Pêchakour. In *les Cahiers du CREAD* N° 72/2005.

[5] Chakour, S.-C., Economie des pêches en Algérie ; Thèse de Doctorat, INA, Alger, Algérie, 2006.

[6] Chakour, S.-C., Une approche empirique pour une gestion durable des pêcheries en Algérie. Colloque international "L'ETAT MALGRE TOUT? ACTEURS PUBLICS ET DEVELOPPEMENT", MONS (Belgique), 14–16 mai 2007.

[7] Gordon, H.S., The economic theory of a common property resource, the fishery, Journal of Political Economics, vol.62, n°2, 1954.

[8] Schaefer, M., Some considerations of population dynamics and economics in relation to the management of marine fisheries, Journal of Fisheries Research Board of Canada, vol.14, n°5, 1957.

[9] Clark, C.-W., Mathematical bioeconomics: The optimal management of renewable resources, Second edition. A Wiley-Interscience Publication, USA., 1931.

[10] Gordon, H.S., An economic approach to the optimum utilization of fisheries resources, Journal of Fisheries Research Board of Canada, vol. 10, n° 7, 1953.

[11] Mesnil, B., Dynamique des populations exploitées et principaux modèles démographiques appliqués à la gestion des pêches; Rapport sur la science et la technologie, Académie des Sciences, rst n° 17, Chapitre 5, décembre, 2003. , Edition TEC & DOC, LAVOISIER, 2003.

Section 5
Environmental performance

Sustainability performance of economic sectors based on thermodynamic indicators

K. J. Ptasinski[1], M. N. Koymans[2] & M. J. C. van der Stelt[1]
[1]*Department of Chemical Engineering, Eindhoven University of Technology, The Netherlands*
[2]*Statistics Netherlands, The Netherlands*

Abstract

Nowadays, various indicators of sustainability performance are used, usually grounded in economics, ecology, thermodynamics, and sociology. The main shortcoming of the commonly used indicators is their 'one-dimensional' character. This paper focuses on thermodynamic sustainability indicators, which also couple environmental and economic aspects. The performance of economic systems is evaluated using various indicators, ranging from exergy, which shows the thermodynamic efficiency, through Cumulative Exergy Consumption CExC, which couples exergy and life cycle analysis, up to Extended Exergy Accounting EEA, where also economic aspects are included. The analysis is illustrated for the Dutch energy sector and for the Dutch Society, including extraction, conversion, agriculture, industry, transportation, tertiary, and domestic sectors.

Keywords: sustainability indicators, exergy analysis, energy policy, environmental impact.

1 Introduction

In the last three decades sustainable development is considered as one of the major components of economic policies in many countries. It is generally believed that improvement in performance of various sectors, particularly industry and transportation, is a very effective way to reduce current global problems due to climate change and environmental pollution. The progress towards sustainability requires meaningful, practical, and scientifically based metrics.

WIT Transactions on Ecology and the Environment, Vol 108, © 2008 WIT Press
www.witpress.com, ISSN 1743-3541 (on-line)
doi:10.2495/EEIA080221

One of the difficulties with measuring sustainability is the lack of consensus on the evaluation of performance of various systems. Sustainability performance is a rather general term and in practice various performance indicators are used, usually grounded in economics, ecology, thermodynamics, and sociology. The main shortcoming of the commonly used sustainability indicators is their 'one-dimensional' character what means a restriction to only one performance aspect. The economic indicators measure the performance in terms of monetary values, such as energy prices. Many environmental indicators are used to express various environmental problems, such as green house effect or acidification. Traditionally, thermodynamic indicators are restricted only to exergetic efficiency, whereas economic and environmental aspects are not involved. A significant problem with complex indicators is that they have to be based on weight factors needed to compare different sustainability aspects and represent the final evaluation in the same units.

The purpose of this paper is to demonstrate how thermodynamic indicators can be coupled with ecological and economic aspects. The paper starts with an explanation of exergy concept, which is the base of all thermodynamic indicators. Two extensions of this concept are described: Cumulative Exergy Consumption (CExC), where exergy is coupled with life cycle analysis, and Extended Exergy Accounting (EEA), where CExC is subsequently combined with other environmental and economic issues. Section 3 demonstrates how the indicators can be applied on the sector level, namely for the analysis of the Dutch energy sector. Finally, in section 4 the application of thermodynamic indicators is illustrated on the national level to analyse the performance of the Dutch society.

2 Thermodynamic sustainability indicators

Energy-related systems are traditionally analysed by balancing the energy and mass flows of a process. The energy efficiency of a process is determined as the ratio between the actual output to the actual input. The energy concept is, however, subject to the First Law of Thermodynamics, which states that energy is conserved in a process. This implies that the output-input ratio always adds up to 100%. However, in the process energy often changes its form and quality, e.g. from mechanical work into heat what is not accounted by the energy analysis.

Exergy analysis is a successor of the traditional energy analysis. This relatively new method has recently been applied in energy and chemical technology, and other fields of engineering and science, Szargut [1]. The exergy concept is based on the Second Law of Thermodynamics, which takes into account not only the quantity of materials and energy flows, but also their quality. Due to the 'entropy law" the quality of materials and energy always degrades in all chemical or physical processes. Exergy is defined as the maximum amount of useful work that can be obtained from a system, and among the different forms of exergy, three forms are the major contributors to the total exergy: thermal exergy, work exergy, and exergy of materials.

Contrary to the mass and energy balances, exergy is not conserved in a process. It means that the exergy leaving any process step will always be less than the exergy in and the difference is called exergy loss or irreversibility I. The exergy balance of a process can be represented in the following form using exergy values of all streams entering and leaving the process:

$$\sum_{IN} \mathrm{E}_j + \mathrm{E}^Q + \mathrm{E}^W = \sum_{OUT} \mathrm{E}_k + I \tag{1}$$

where $\sum_{IN} \mathrm{E}_j$ and $\sum_{OUT} \mathrm{E}_k$ are exergy flow of all entering and leaving material streams, respectively, E^Q and E^W are the sums of all thermal exergy and work interactions involved in a process. Exergy loss (irreversibility) relates to entropy production in the system.

Exergy-based indicators have been coupled with Life Cycle analysis concepts to become a 'two-dimensional' indicator, such as Cumulative Exergy Consumption (CExC), as proposed by Szargut [1]. The CExC method values a product based on its entire life from cradle-to-grave and can be applied for production-chain analysis. In this method the net of production process is divided in four levels, comprising all material and energy streams involved in the final production process (level 1), and in the fabrication of intermediate products (level 2), as well in production of machines and installations (levels 3 and 4). The CExC-value indicates the total amount of exergy consumed per final product in the whole production chain.

CExC and exergy methods produce results of a purely technical nature that are both expressed in the same units – joules. The results of exergy evaluation are often not directly suitable for non-technical analysis where the monetary values are preferred. Recently, Extended Exergy Accounting method (EEA) has been proposed as an extension of standard exergy analysis to include also economic and environmental issues, Sciubba [2]. The extension of the classical exergy concept by capital and labor equivalents has been motivated by Sciubba by Neo-Classical Economics (NCE) where capital and labor are identified as production factors that contribute to performance. The advantage of performance indicators based on the EEA is that they can be used as exergetic as well as monetary metrics for all stages of production processes.

The Extended Exergy contains the following parts: feedstock exergy FE being the CExC of feedstock, capital equivalent exergy CEE, labour equivalent exergy LEE, and environmental remediation exergy ERE. The Extended Exergy can be calculated from its constituent's components:

$$\mathrm{EE} = \mathrm{FE} + \mathrm{CEE} + \mathrm{LEE} + \mathrm{ERE} \tag{2}$$

The FE component can be calculated based on mass-flow data. In order to address the economic (CEE and LEE) and environmental (ERE) components production statistics of a sector are required. Moreover, the capital conversion factor, specific for the analysed system, has to be evaluated to express the monetary value of a certain quantity of exergy.

3 Performance of the Dutch energy sector

Methodological reasons to study the energy sector are large volumes of natural resources processed, thereby directly being relevant to environmental issues such as natural resource scarcity and atmospheric pollution. The analysis of the Dutch energy sector 1996 is based on mass-flow data published by the Statistics Netherlands [3]. The data allows for a breakdown into 8 branches, which are classified in three sub-sectors, as indicated below:

- *Exploitation*
- *Transformation*: cokeries; refineries; central electricity & heat production; decentral electricity & heat production; refuse incinerators
- *Distribution*: solid fuel trade; oil product trade; distribution of water, gas, electricity and heat.

For the above-mentioned branches, the 27 mass-flow accounts of primary resources are grouped into three main categories: hard coal (products), crude oil (products), and other energy carriers (electricity, steam).

Table 1 shows the main components of the Dutch energy balance, Ptasinski *et al* [4]. The energy balance is dominated by trade as the total amounts of trade (import and export) exceed the domestic exploitation and consumption. Both import and export products are mainly crude oil and oil products. Table 2 summarises the energy, exergy and CExC flows for all branches of the Dutch energy sector, calculated using the annual mass-flow data of every of the above-mentioned 27 resource account, and its net-calorific value, its specific exergy value, and a value representing its Cumulative Exergy Consumption (CExC), respectively.

Table 3 shows three efficiency indicators (η), defined as the ratio of production (output) and input, based on the three different valuations from the Table 2. For the entire sector, the energy η_{EN} and exergy η_{EX} based indicators are almost the same. This is due to the fact that the net-calorific values and specific exergy values for all present substances are quite the same. The CExC-indicator η_{CExC} shows significant lower values, reflecting a 'cradle-to-gate' history of a feedstock and involving a partial life cycle analysis. Comparing the different

Table 1: Energy balance 1996 for the Netherlands.

Input (PJ)			Output (PJ)		
Exploitation		3 119	Domestic consumption		3 076
- *natural gas*	*2 891*		- *crude oil*	*2 441*	
Import		6 506	- *natural gas*	*1 598*	
- *crude oil*	*4 227*		- *crude oil products*[a]	*-1737*	
- *crude oil products*	*1 356*		Export		5 960
Bunker		- 601	- *crude oil*	*1 898*	
Stock mutation		12	- *crude oil products*	*2 485*	
			- *natural gas*	*1 464*	
Total		9 036	Total		9 036

[a] The negative output in this case means production exceeds consumption.

sub-sectors, it is noted that where the activity does not involve transformation of the feedstock, the η_{EN} and η_{EX} indicators approach unity. The transformation sub-sector shows lower indicators, which is explained by a lower specific exergy value of the product compared to the feedstock.

www.witpress.com, ISSN 1743-3541 (on-line)

Table 2: Energy and exergy flows for sub-sectors of the Dutch energy sector.

Sub-sector	Input (PJ)			Production (PJ)	
	Energy	Exergy	CExC	Energy	Exergy
Exploitation	3 202	3 337	3 372	3 169	3 303
Transformation	5 647	5 973	6 217	5 116	5 407
Distribution	5 625	5 942	6 649	5 595	5 904
Dutch Energy Sector	14,474	15,252	16,238	13,879	14,615

Table 3: Efficiencies for sub-sectors of the Dutch energy sector.

Sub sector	η_{EN}	η_{EX}	η_{CExC}
Exploitation	0.99	0.99	0.98
Transformation	0.91	0.90	0.87
Distribution	0.99	0.98	0.89
Dutch Energy Sector	0.96	0.96	0.90

The EEA is performed for the following four branches within the Dutch energy sector, where the production and monetary statistics are available: 1. cokeries and refineries, 2. refineries, 3. central electricity production, 4. distribution and decentral electricity production. The first three branches are part of the transformation sector and the fourth relates to two sub-sectors: transportation and distribution. Table 4 summarizes the energy and exergy data for these branches.

Table 4: Energy and exergy flows for branches within the Dutch energy sector.

Branch	Input (PJ)		Output (PJ)	
	Energy	Exergy	Energy	Exergy
1 cokeries & refineries	4 978	5 334	4 786	5 133
2 refineries	4 856	5 221	4 682	5 022
3 central electricity production	526	511	232	216
4 distribution & decentral electricity production	1 315	1 228	1 265	1 184

The components of extended exergy can be expressed both as the exergetic as well as monetary values using the capital conversion factors. The specific capital conversion factor K_{cap} has been calculated as

$$K_{cap} = FE^{€} / CExC^{FE} \tag{3}$$

where $FE^{€}$ is the annual monetary value of the feedstock, and $CExC^{FE}$ represents the annual cumulative exergy value of the feedstock. Table 5 shows the $CExC^{FE}$, $FE^{€}$, and specific K_{cap} values for the four branches. The branch-specific capital conversion factors K_{cap} are used to estimate the CEE, LEE and ERE components of the extended exergy, assuming this conversion factor to be branch specific but the same for all EE-terms. The capital equivalent exergy CEE is estimated

Table 5: Capital conversion factors for branches within the Dutch energy sector.

Branch	$CExC^{FE}$ (PJ)	$FE^{€}$ (mln €)	K_{cap} (mln €/PJ)
1. cokeries & refineries	5 484	8 929	1.63
2. refineries	5 367	8 650	1.61
3. central electricity production	564	1 309	2.32
4. distribution & decentral electricity production	1 728	8 759	5.07

by conversion of the monetary values of short and long-term investments. The component labour equivalent exergy LEE has three contributions: the man-power equivalent exergy, labour equivalent exergy due to skills, and social accounts. The Environmental Remediation Exergy ERE is defined as the exergy consumption required to neutralize the impact of waste flows entering the environment. As this section applies EEA on sector (company) level, a remediation process to neutralize the annual waste flows of these companies together cannot be designed. The ERE component is calculated only for branch 2, refineries, where within the production statistics the environmental costs are available.

Table 6 shows the values of all terms contributing to the extended exergy for all considered branches. Branches 1 and 2, cokeries and refineries, are dominated by the feedstock term, followed by capital, and labour, respectively. The large contribution of feedstock to the overall EE can be seen as indicator for the pressure of the production system on the environment, and the dependence of the system on the environment. The branches 3 and 4 present a quite other image. Central electricity and heat production (branch 3) depends far less on feedstock (mainly natural gas) and capital goods makes up for 80% of the EE of the product, reflecting the capital-intensive character of the branch.

The performance indicator of considered energy branches can be represented as

$$\eta_{EE} = Ex_p / EE \tag{4}$$

where Ex_p and EE are the annual chemical and extended exergy values, respectively, for every branch. Figure 1 shows a comparison of efficiencies for all branches based on cumulative exergy consumption and extended exergy, respectively. The η_{EE} indicator is much lower than η_{CExC}, but also far more strict.

Table 6: Extended exergy and its constituent components (PJ).

Branch	FE	CEE	LEE	ERE	EE
1 cokeries & refineries	5 484	2 994	459		8 937
2 refineries	5 367	2 846	501	4	8 718
3 central electricity production	564	2 829	153		3 546
4 distribution & decentral electricity production	1 728	2 126	266		4 120

It can be concluded that the thermodynamic performance of refineries (branch 2) is very good; $\eta_{CExC} = 94\%$, but the performance of the branch taking into account capital and labour terms, is substantially lower; $\eta_{EE} = 58\%$.

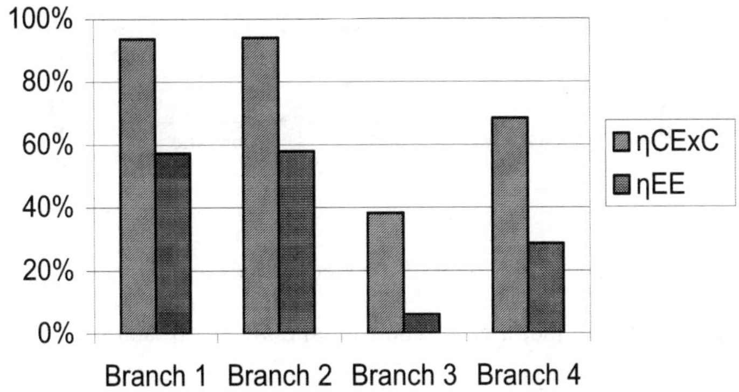

Figure 1: CExC and EE exergy efficiency for different energy branches.

4 Analysis of the Dutch society

In the last decades a number of society exergy analysis have been performed and recently the EEA has been applied for the analysis of the Italian, Milia and Sciubba [5], and Norwegian society, Ertesvåg [6]. In this section the Dutch society is analysed using energy and exergy based indicators, including EEA. To this end the country – the Netherlands as the system is divided into the following sectors: Ex–extraction (extraction of resources and raw materials from the environment), Co–conversion (energy conversion systems, including heat and power plants), Ag–agriculture (farming, herding and fishing activities, including related industry), In–industry (manufacturing industries except food industry and oil refineries), Tr–transportation (transportation of peoples and goods), Te–tertiary sector (services including government, banks, schools, commerce), and Do–domestic sector (households), as shown in Figure 2.

The system is surrounded by: E–Environment (the Earth crust, the atmosphere, the oceans, etc), and A–Abroad (other countries). All arrows between these sectors represent fluxes, which are classified as: R–primary resources (primary: fossil fuels, metals, minerals, and secondary resources: from petroleum refining, and electrical energy), N–natural resources (agricultural products, wood, livestock, fish, game), P–products (products generated by In, Tr, and Te–sectors), T–trash (waste materials deposited in the environment), D–discharge (combustion gases, thermal discharge in the environment), L–human work (labour), and C–capital (monetary flows).

The evaluation of the performance of the Dutch society in 2000 using exergy indicators is based on energy balances published by the Statistics Netherlands [2], where data on R, N, P, T, and D-fluxes are included. The capital equivalent exergy CEE (see Eq. 2) is calculated as the product of a monetary $C^{€}$-flux and

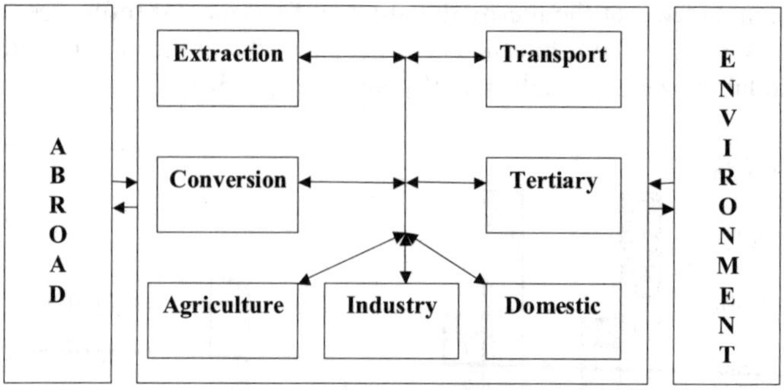

Figure 2: Model of the Netherlands as the system.

the capital conversion factor K_{cap}. The capital conversion factor K_{cap} is evaluated as

$$K_{cap} = E_{in} / M_2 \tag{5}$$

using E_{in} as the annual exergetic input to the society, and M_2 as the monetary circulation in the country. The labour equivalent exergy (see LEE in Eq. 2), being the exergetic equivalent of a monetary $L^{€}$-flux, is calculated as the product of a monetary $L^{€}$-flux, and the labour conversion factor K_{lab}. The labour conversion factor K_{lab} is evaluated as

$$K_{lab} = E_{in} / W_{tot} \tag{6}$$

using E_{in}, W_{sector} and W_{tot} as the total amount of work-hours in the sector and society, respectively. The capital flux into each sector (except the households) is taken as the sum of production, the gross investment, and net production subsidies. The capital flow out of the sectors is taken as the sum of the cost of good and services consumed in the production, compensation of employees, net product taxes, return to the owners, and gross investment. Labour is considered as an output from the household Do-sector and input to all other sectors.

In 2000, the values of E_{in}, M_2, and W_{tot} for the Dutch society are 5490 PJ, 353 billion €, and 9843 Mhr, respectively. The calculated capital and labour conversion factors for the Netherlands are: K_{cap} = 15.7 MJ/€ and K_{lab} = 557.8 MJ/hr, van der Stelt *et al* [7].

Table 7 summarizes the EE fluxes to all sectors of the Dutch society whereas Table 8 shows the conversion efficiency of individual sectors, calculated as the ratio between output to a sector including R, P, N, L, and C-fluxes, and corresponding input. The energy and exergy efficiency are very similar for every sector due to the way of accounting fluxes in this method.

The main fluxes are resources and products for which the energy and exergy values do not differ substantially.

The EE efficiency is notably larger than the exergy efficiency for almost all sectors (excluding industry) what is due to large contribution of the capital outflow with respect to capital inflow to a sector. The EE efficiency in the extraction sector is almost the same as the exergy efficiency what indicates less influence of the capital and labour factors in this sector. The extraction, conversion, and industry sectors exhibit the highest efficiency, mainly to a large throughput of exergy of resources and products. The EE efficiency of the agriculture is higher than its exergetic counterpart due to relatively high salaries in this sector. A similar situation can be observed for the transportation sector. The performance of the household Do-sector is different from other sectors what is quite characteristic in the EEA method. Within the Do-sector the extended exergy can be produced due to capital increase what can be regarded as a result of man's creativeness. From this point of view the second law of thermodynamics is not violated, as pointed by Ertesvåg [6].

Table 7: Extended exergy fluxes of Dutch sectors (PJ/year).

	Ex	Co	Ag	In	Tr	Te	Do
Input							
R	7792.3	6239.2	344.3	974.1	543.7	275.6	564.5
P	8.6	25.1	110.6	869.7	9.0	694.0	276.3
N			808.3			64.6	58.1
L	8.4	39.0	212.0	769.2	394.9	4066.8	0.0
C	194.4	563.4	1030.9	2460.6	901.1	7589.3	7154.3
Output							
R	7558.1	4974.7					
P	2.1	17.5		1328.8	296.8	238.9	41.4
N			320.6			58.1	
L							5490.2
C	218.3	680.0	1102.2	2428.9	971.4	7483.9	7223.6
T	1.3	0.1		13.8			8.1
D	36.6	319.3	17.7	80.0	28.2	7.0	37.2

Table 8: Conversion efficiencies in various sectors of the Dutch society (%).

	Ex	Co	Ag	In	Tr	Te	Do
Energy	96.8	71.9	26.6	84.9	62.8	29.7	6.0
Exergy	96.0	71.0	26.8	84.9	57.8	30.4	6.0
Extended exergy	97.1	82.6	56.7	74.1	68.6	61.3	160.0

5 Conclusion and discussion

The analysis of the Dutch energy sector shows that energy and exergy efficiencies of most sub-sectors are very high but the CExC efficiency is lower, accounting for the exergy consumption of the feedstock before it enters the sector. The EE efficiencies are much lower than those using energy, exergy or CExC based indicators as economic and environmental aspects are taken into account. The EEA of the Dutch society shows that the exergetic equivalents of capital and labour have also large influence on the efficiency in comparison with energy and exergy efficiencies, especially in the tertiary and domestic sectors. The production of EE in the Do-sector can be seen as a product of man's creativity. In all sectors, with the exception of the industry, the EE efficiency is higher than the energy and exergy efficiency. This is due to the lower capital conversion factor of the whole society, 15.7 MJ/€ as evaluated in section 4, compared to the average value of 470.4 MJ/€ for the energy sector as evaluated in section 3. The energy sector is more resource intensive and therefore the ratio of exergetic to monetary fluxes for this sector is higher that for the whole society. EEA is grounded through exergy concept in thermodynamics and on the other hand, through production factors capital and labour, in economics. Moreover, environmental costs are included in this analysis. Therefore EEA seems to be a proper candidate to be used as a multidimensional indicator to analyse performance of chemical and energy transformations.

References

[1] Szargut, J., *Exergy Method, Technical and Ecological Applications*, WIT Press: Southampton, Boston, 2005.

[2] Sciubba E., Cost analysis of energy conversion systems via a novel resource-based quantifier. *Energy,* **28**, pp. 457–477, 2003.

[3] Statline. The electronic data bank of Statistics Netherlands, http://statline.cbs.nl.

[4] Ptasinski, K.J., Koymans M.N. & Verspagen H.H.G., Performance of the Dutch energy sector based on energy, exergy and extended exergy accounting. *Energy*, **31**, pp. 3135–3144, 2006.

[5] Milia, D. & Sciubba E., Exergy-based lumped simulation of complex systems: an interactive analysis tool. *Energy,* **31**, pp. 100–111, 2006.

[6] Ertesvåg, I.S., Energy, exergy, and extended-exergy analysis of the Norwegian society 2000. *Energy*, **30**, pp. 649–675, 2005.

[7] Van der Stelt, M.J.C., Koymans M.J.N., Sciubba, E. & Ptasinski, K.J., Performance of the Dutch society based on energy, exergy, and extended exergy accounting. *Proc. of the 20th Int. Conf. on Efficiency, Cost, Optimization, Simulation and Environmental Impact of Energy Systems ECOS 2007*, eds E. Mirandola, Ö. Arnas & A. Lazzaretto: Padova, pp. 1477–1484, 2007.

The influence of regional autonomist government on the territory environmental and economic performances

G. Skonieczny & B. Torrisi
Department of Economics and Territory,
Branch of Economic Statistics, University of Catania, Italy

Abstract

In the last decade there's been a growing interest in the Regions with special autonomy, in particular from a political point of view, and less from a scientific one.

In Italy, some Regions were given special autonomy after the coming into force of the Republican Constitution. In Spain, such regions are called "Comunidades Autònomas". They are similar to the Italian Regions, but with big differences concerning their provisions and power to make decisions.

The few scientific contributions available have analysed these territorial structures mainly from a normative and managerial point of view. Some sporadic attempts at economic analysis have been made, but always from a merely theoretical point of view.

In this context, there seems to be a lot of space for the development of a trend of research aiming at a quantitative analysis of these regions, in comparison with less autonomist areas. Such quantitative investigations on the competitiveness, socio-economic development, quality of life, social welfare, and environmental performance of these areas with special autonomy, are issues still open and not entirely analysed.

As a consequence of this void in the research, the present contribution aims at analysing how much influence the autonomy of a territory can have on the changes in its performances. We've taken into consideration some Italian and Spanish regions (both with special autonomy and not).

In order to attain this objective, we have used those statistical indicators that can synthesise the above mentioned aspects and point out the similarities and dissimilarities among the considered areas between 1995 and 2006. From the methodological point of view, we have used the multi-varied analysis techniques, in order to analyse the territorial homogeneities and heterogeneities, and the parametric and non-parametric tests, to verify whether, some conditions being equal, the autonomist aspect can produce changes in the performances of the analysed territories.

Keywords: regions with special autonomy, classification, graduation, territorial comparisons.

WIT Transactions on Ecology and the Environment, Vol 108, © 2008 WIT Press
www.witpress.com, ISSN 1743-3541 (on-line)
doi:10.2495/EEIA080231

1 Introduction

The competitiveness of an area, its economic, social, and environmental development, its quality of life and welfare, represent the main attractive aspects of a territory.

In this context, it seems possible to assume that more autonomous systems of a territory government, than the standard of the country it belongs to, can create more wealth, more control of the territory, better environmental development, and therefore, better conditions of life.

There are few scientific contributions aiming at verifying such thesis; they are mainly theoretical, not empirical.

As a consequence of this void in the research, the present contribution aims at analysing how much influence can the autonomy of a territory have on the changes in its performances.

In order to attain this objective, we have used those statistical indicators that can synthesise the above mentioned aspects and point out the similarities and dissimilarities among the considered areas between 1995 and 2006. Moreover, the graduation techniques and the non-parametric statistical tests have generated the conclusive analyses of this work.

2 The sources, the data, and the indicators

The starting point of this analysis is the collection of data and the selection of the indicators used. The latter include:

- ***Environment***

Public water provided per capita - Rubbish collection per capita - % of rubbish going to final distribution

- ***Economic Aggregates***

Gross value added to basic prices per branch - A_B Agriculture, hunting, forestry and fishing - C_E Mining and quarrying, manufacturing and electricity - F Construction - G_H_I Wholesale and retail trade, repair of motor vehicles, motorcycles and personal and household goods; hotels and restaurants; transport, storage and communication - J_K Financial intermediation; real estate, renting and business activities - L_to_P Public administration and defence, compulsory social security - education; health and social work; other community, social and personal service activities; private households with employed persons

- ***GDP***

GDP in SPA per inhabitant - Average growth of regional GDP in percentage

- ***Family Accounts***

Per capita family income, in PPCS - Family income as percentage of GDP -

- ***Employment and Unemployment***

Persons in employment – men (1000)/Male pop. Age 15-64 - Persons in employment – women (1000)/Female pop. Age 15-64 - Persons in employment – full time (1000)/Persons in employment – total - Persons in employment – part time (1000)/Persons in employment – total

- ***Unemployment***

Total – Female - <25 years old - Long term unemployment as % of unemployment. - Trend in unemployment rate, average annual increase (%) - Trend for unemployed people, average annual increase (%) - Trend in work-force, average annual increase (%)

- ***Population***

Born (°/oo) - Dead (°/oo) - Net migration (°/oo) - Change of population (°/oo) - Density of population - Birth-rate - Death-rate per 1000 inhabitants, 1000 - Population natural increase rate - Young age dependency ratio (0 – 19)(20 – 59) - Old age dependency ratio (60+)(20 – 59)

- ***Transport***

Railways (km) total/territory - Railways with more than double quantity of lines (km)/tot railways - Railways – electrical lines/tot railways - Motorways (km)/territory - Other roads (Km)/territory - Number of deaths per million of inhabitants

- ***Tourism***

Establishments/total establishments - Bedspace/establishments - Arrivals of residents /tot arrivals - Arrivals of non-residents/tot arrivals - Nights spent by residents/tot nights - Nights spent by non-residents/tot nights.

The data used have been taken from the classification of European regional areas NUTS 2 (European Commission Regions, 2003 and 2007 – Atkinson et al. 2001, Guarini 1998). In particular, we will refer to 208 regions of the EU-15 for 1995, and to 343 regions of the EU-27 for 2006, classified on the base of 48 indicators (variables).

The first selection of variables has been determined from the uniformity and completeness of the data matrixes in comparison with the considered variables. Some indicators have been cancelled because of the series incompleteness (75 out of 123). In all, we have created a data base of 208 regions (year 1995) and 343 regions (year 2006) for 48 referred variables expressing environmental, social, and economic development.

The use of such indicators has provided the first information on the environmental, social, and economic development of the analysed territory.

3 The methodology employed

The 208 and 343 regions have been classified through the metric procedure of classification *k-means*. (This procedure allows one to identify groups of relatively homogeneous cases in relation with the selected features, using an algorithm which can handle a great number of cases. Such algorithm needs the number of clusters.) In relation to the 48 indicators (as variables effect) chosen and selected with the principal components method. The cluster analysis allowed us to aggregate the homogeneities of the analysed territories in 5 clusters.

Since the simple aggregation has resulted in a little significant representation, in order to define a list of distances between one territorial unit and the others, it was necessary to build an algorithm GDM (*Graduate Distance from Matrix*). In particular, the procedure adopted to generate a comprehensive measure d_i of the effect induced by environmental and economic development (where i represents the ith territorial unit), was based on the remodulation of some indicators according to the reciprocal calculus, as to generate a uniform datum reading for all the variables used. Then we standardised the variables in relation to the mean and the mean-square deviation. Moreover, we applied the PLS methods to the initially selected variables, in order to generate the loadings p_j which have been used in the composition of the final indicator for each jth variable.

The indicator we built is given by

$$d_i = \sqrt{\sum_{i=1}^{n}\sum_{j=1}^{m}(x_{ij} - \min_j)^2 p_j}$$

where x_i represents the standardised values in comparison with the jth variable in relation to the jth territorial unit, while min_j represents the minimum value of the distribution associated to the jth variable.

The use of such indicator has allowed us to provide an assessment of the degree of environmental and economic development. Through d_i we could generate a comprehensive classification of the 208 (1995) and 343 (2006) regional areas of European countries. Being the mere reading of the list quite simplistic in comparison with the initial objectives, we tried to represent the 5 clusters of the European regions according to the estimated level of the environmental and economic development. Assuming a scale of development degree divided into 5 classes (low, medium-low, medium, medium-high, high), we built a graphic representation of the quantities on the base of the following function:

$$\beta = d_{i*}\, \sigma_k$$

where β represents the distance from the centre (point of minimum estimated economic and environmental development), d_i an increasing measure in comparison with the level of the environmental and economic development in the ith region, and σ_k an also increasing scale factor, while the angular coefficients have been considered uniform. The scale factor σ_k represents the mean-square deviations between the territorial units of d_i in comparison with the mean M_{dk} :

$$\sigma_k = \sqrt{\frac{1}{k}\sum_{i=1}^{k}(d_i - M_{dk})^2}$$

where M_{dk} represents the arithmetic mean of the levels d_i for each $i=1...n$ and $k=2...n-1$. By using such criterion, it was possible to represent the classification of the 208 regional areas on the base of the environmental and economic development, estimated d_i within each class of level, which could extrapolate the closest distances in comparison with the furthest, starting from the matrix of distances and similarities. In this way, it was possible to build a mapping of distances, and to verify how far is an ith region from the nth, pointing out the position of areas with special autonomy in such context (*see fig. 1 and 2*).

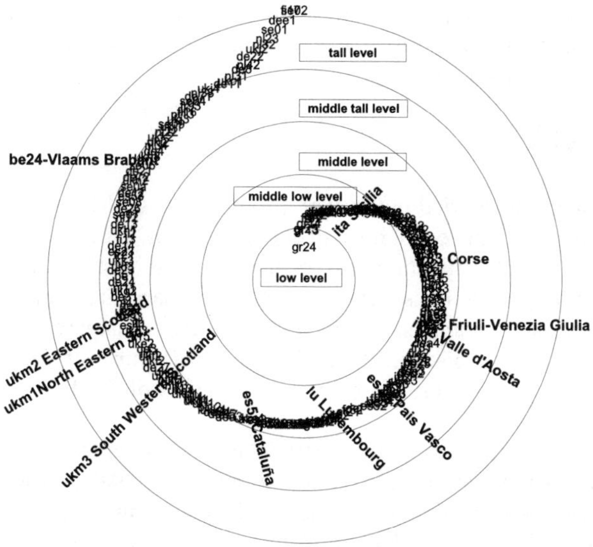

Figure 1: Classification of the 208 regions EU-15 (year 1995) per levels of environmental and economic development.

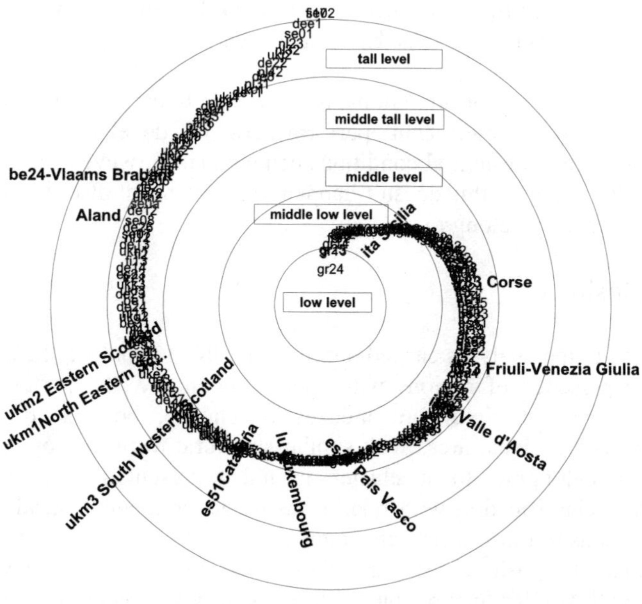

Figure 2: Classification of the 343 regions EU-27 (year 2006) per levels of environmental and economic development.

In this way, it was possible to point out the levels of the considered development in single autonomous regions, and their distance from the minimum and maximum estimated level.

4 The results

Through the analyses carried out, it was possible for us to define the measure of the environmental and economic development. In particular, through the methodologies of classification we generated 5 groups which provided a comprehensive picture of the similarities and dissimilarities among the regions in relation to the economic and environmental indicators considered (*see tab. 1*).

In this way, it was possible to define the positioning of the regions with special autonomy in comparison to the other regions of the European countries within each cluster.

The analysis we adopted and the particular graphic representation allowed a better reading of the state of economic and environmental development of the analysed regions, with particular attention to the positioning of those with special autonomy and to the change which took place between 1995 and 2006 (*see fig. 1 and 2*).

By comparing the regions with special autonomy to other regions with similar numerousness of population, it was possible to point out that between 1995 and 2006 the most significant changes were recorded in the first ones. The application of the Wilcoxon test has established the statistical significance of such differences (*see tab. 2*).

On the base of such considerations, it is possible to confirm that autonomous regions make more significant performances of development than non-autonomous regions, structural conditions being equal. However, in certain cases (like in Sicily or Corse) this doesn't happen. But this result also depends on the national context they belong to.

5 Conclusions

With relation to the analyses carried out, it is possible to confirm the assumption that, in the presence of regions with special autonomy, the performances of development present significant indexes of change, both in time and in comparison with regional areas with similar structural features (population). We proved that belonging to a cluster is not representative of the region performances classification with relation to its government. Instead, we found that comparisons in the two lists are significant (most of the autonomous regions present quite good positioning), particularly those comparisons between couples of regions with similar features but with kinds of government and indicators of development that are, in some cases, very different.

The possibility to interpret this phenomenon and the simple reading of the results are encouraging to carry out other studies, hoping that the matrixes of the data and the indicators will be complete.

Table 1: Results of the cluster analysis and regions with special autonomy in bold.

Cluster 1
fr91 Guadeloupe (FR) - fr92 Martinique (FR) - fr93 French Guiana (FR)- ie01 Border, Midlands and Western - fr94 Reunion (FR)
Cluster 2
be25 West-Vlaanderen - be31 Brabant Wallon - be32 Hainaut - be33 Liège - deb1 Koblenz - dee3 Magdeburg - def Schleswig-Holstein - deg Thüringen - es22 Comunidad Foral de Navarra - es23 La Rioja - es24 Aragón - es3 Comunidad de Madrid - fr63 Limousin - fr72 Auvergne - it11 Piemonte - nl11 Groningen - fi16 Uusimaa (suuralue) - fi2 Åland - se01 Stockholm - ukc2 Northumberland, Tyne and Wear - ukj2 Surrey, East and West Sussex - ukj3 Hampshire and Isle of Wight - ukj4 Kent - ukk1 Gloucestershire, Wiltshire and North Somerset - de14 Tübingen - de21 Oberbayern - de72 Gießen - de73 Kassel - gr22 Ionia Nisia - gr23 Dytiki Ellada - gr24 Sterea Ellada - gr25 Peloponnisos - es53 Illes Balears - es61 Andalucia - es62 Murcia - es63 Ceuta y Melilla (ES) - nl31 Utrecht - nl32 Noord-Holland - nl33 Zuid-Holland - nl34 Zeeland - uke2 North Yorkshire - uke4 West Yorkshire - ukf1 Derbyshire and Nottinghamshire - ukf2 Leicestershire, Rutland and Northants - **ukm1 North Eastern Scotland - ukm2 Eastern Scotland - km3 South Western Scotland** - ukm4 Highlands and Islands - fr53 Poitou-Charentes - fr62 Midi-Pyrénées - fi14 Väli-Suomi - fi15 Pohjois-Suomi - be35 Namur - de11 Stuttgart - de12 Karlsruhe - de13 Freiburg - gr12 Kentriki Makedonia - gr13 Dytiki Makedonia - gr14 Thessalia - gr21 Ipeiros - es42 Castilla-la Mancha - es43 Extremadura - **es51 Cataluña** - es52 Comunidad Valenciana - nl13 Drenthe - nl21 Overijssel - nl22 Gelderland - nl23 Flevoland - ukk3 Cornwall and Isles of Scilly - ukk4 Devon - ukl1 West Wales and The Valleys - ukl2 East Wales - de94 Weser-Ems - dea1 Düsseldorf - dea2 Köln - dea3 Münster - gr41 Voreio Ai gaio - gr42 Notio Ai gaio - gr43 Kriti - es11 Galicia - es12 Principado de Asturias - es13 Cantabria - ukg3 West Midlands - uke1 East Riding and North Lincolnshire - ukh3 Essex - ukj1 Berkshire, Bucks and Oxfordshire - fi13 Itä-Suomi - ukh1 East Anglia - ukd4 Lancashire - ukd5 Merseyside - fr51 Pays de la Loire - fr52 Bretagne - be34 Luxembourg (B) - gr11 Anatoliki Makedonia, Thraki - es41 Castilla y León - nl12 Friesland - ukd1 Cumbria - ukk2 Dorset and Somerset - de8 Mecklenburg-Vorpommern - gr3 Attiki - es7 Canarias (ES) - nl41 Noord-Brabant - ukf3 Lincolnshire - ukn Northern Ireland - ukh2 Bedfordshire, Hertfordshire - ukd3 Greater Manchester - fr1 Île de France - pt12 Centro (PT) - ukg1 Herefordshire, Worcestershire and Warks - dea5 Arnsberg - fr21 Champagne-Ardenne - pt13 Lisboa e Vale do Tejo - ukg2 Shropshire and Staffordshire - **es21 Pais Vasco** - pt14 Alentejo - **be24 Vlaams Brabant - it12 Valle d'Aosta - - it33 Friuli-Venezia Giulia**
Cluster 3
at21 Kärnten - at22 Steiermark - at31 Oberösterreich - at33 Tirol - uke3 South Yorkshire - uki1 Inner London - uki2 Outer London - at32 Salzburg - ie02 Southern and Eastern - at11 Burgenland - at12 Niederösterreich - at13 Vienna - at34 Vorarlberg - pt2 Açores (PT) - pt3 Madeira (PT) - ukd2 Cheshire
Cluster 4
be21 Antwerpen - be22 Limburg (B) - be23 Oost-Vlaanderen - - fr24 Centre - fr25 Basse-Normandie - fr26 Bourgogne - fr3 Nord - Pas-de-Calais - it52 Umbria - it53 Marche - it6 Lazio - it71 Abruzzo - de3 Berlin - de4 Brandenburg - de5 Bremen - de6 Hamburg - fr81 Languedoc-Roussillon - fr82 Provence-Alpes-Côte d'Azur - **fr83 Corse** - - **ita Sicilia** - itb Sardegna - lu Luxembourg - pt11 Norte - de23 Oberpfalz - de24 Oberfranken - de25 Mittelfranken - de26 Unterfranken - fr41 Lorraine - fr42 Alsace - fr43 Franche-Comté - fr61 Aquitaine - it72 Molise - it8 Campania - it91 Puglia - it92 Basilicata - de71 Darmstadt - de91 Braunschweig - de92 Hannover - de93 Lüneburg - it13 Liguria - it2 Lombardia - it31 Trentino-Alto Adige - it32 Veneto - pt15 Algarve - fr23 Haute-Normandie - it4 Emilia-Romagna - it51 Toscana - de27 Schwaben - fr71 Rhône-Alpes - it93 Calabria - fr22 Picardie
Cluster 5
de22 Niederbayern - dea4 Detmold - deb2 Trier - deb3 Rheinhessen-Pfalz - nl42 Limburg (NL) - fi17 Etelä-Suomi - se02 Östra Mellansverige - se04 Sydsverige - ded1 Chemnitz - ded2 Dresden - ded3 Leipzig - dee1 Dessau - se07 Mellersta Norrland - se08 Övre Norrland - se09 Småland med öarna - se0a Västsverige - dec Saarland - se06 Norra Mellansverige - dee2 Halle - ukc1 Tees Valley and Durham

Table 2: Results of p-value with the Wilcoxon Test on the delta 1995-2006 among regions with similar numerousness of population in relation to the environmental and economic variables - *non-significant test for p value > 0,05.

	Transport	Tourism	Environment	Economic aggregates
Prov. Vlaams Brabant Prov. Liège	0,023	0,0012	0,012	0,027
Pais Vasco Castilla y León	0,03	0,005	0,034	0,049
Cataluña Andalucia	0,076	0,041	0,023	0,038
Corse Martinique (FR)	0,89*	0,0034	0,54*	0,555*
Valle d'Aosta Molise	0,02	0,004	0,0005	0,005
Friuli-Venezia Giulia Abruzzo	0,031	0,005	0,05	0,0075
Sicily Lazio	0,78*	0,745*	0,56*	0,575*
Malta Luxembourg (Grand-Duché)	0,026	0,0042	0,015	0,03
Åland Pohjois-Suomi	0,0056	0,021	0,034	0,049
North Eastern Scotland Lincolnshire	0,045	0,01	0,023	0,038
Eastern Scotland West Wales and The Valleys	0,048	0,013	0,034	0,049
South Western Scotland Gloucestershire, Wiltshire and N.S.	0,0057	0,022	0,033	0,048
	GDP	**Family accounts**	**Employment**	**Unemployment**
Prov. Vlaams Brabant Prov. Liège	0,022	0,017	0,012	0,007
Pais Vasco Castilla y León	0,044	0,039	0,034	0,029
Cataluña Andalucia	0,033	0,028	0,023	0,018
Corse Martinique (FR)	0,55*	0,545*	0,54*	0,535*
Valle d'Aosta Molise	0,01	0,005	0,50*	0,495*
Friuli-Venezia Giulia Abruzzo	0,007	0,065	0,06	0,055
Sicily Lazio	0,587	0,567	0,456	0,235
Malta Luxembourg (Grand-Duché)	0,025	0,02	0,015	0,01
Åland Pohjois-Suomi	0,044	0,039	0,034	0,029
North Eastern Scotland Lincolnshire	0,033	0,028	0,023	0,018
Eastern Scotland West Wales and The Valleys	0,044	0,039	0,034	0,029
South Western Scotland Gloucestershire, Wiltshire and North Somerset	0,043	0,038	0,033	0,028

References

[1] Atkinson A.B., Cantillon B., Marlier E. and Nolan B. (2001), *Indicators for social inclusion in the European Union*, report presented at Conference on Indicator for social inclusion: making common EU, Antwerp, 14–15 September.
[2] European Commission, Regions: Statistical Yearbook 2003–2007, Lussemburgo.
[3] Horn R.V. (1980), *Social indicators: meaning, methods and applications*, in International Journal of Social Economics vol.7 n.8 pp.421–460.
[4] Marriem I.C. (1968), *Welfare and its measurement*, in indicators of social change concept & measurement ed. By Elanor Berset Sheldon – New York.
[5] Skonieczny G. and Torrisi B (2004) "*La misura del benessere sociale nelle regioni europee*" in Rivista Italiana di Economia Demografia e Statistica (Gennaio Giugno).

How to find the most practical ecosystem management plan

T. C. Haas
Lubar School of Business Administration,
University of Wisconsin at Milwaukee, USA

Abstract

A predictive understanding of how political processes produce a sequence of ecosystem management decisions would allow environmental managers to estimate how much social change would be needed to make an ecosystem sustainable. To this end, a stochastic, temporal model of how political processes influence and are influenced by ecosystem processes is being developed. This model is realized in a set of interacting Influence Diagrams (Bayes Nets with Decision nodes) that each represent the belief systems of political groups in countries that affect an ecosystem. These group models also interact with a model of the affected ecosystem.

After these models are fitted to a political-ecological data set, they are used to find the most practical ecosystem management plan by modifying the modeled group belief systems that were estimated from the data, until a sequence of group actions towards the ecosystem over a future time period results in desired ecosystem values at a designated future time point, e.g. a viable wildlife population in the year 2058. Belief systems are minimally modified away from their data-based values so as to produce this desired sequence of group actions towards the ecosystem.

Such a set of interacting models has been constructed for the management of the endangered Cheetah (*Acinonyx jubatus*) across Kenya, Tanzania, and Uganda. Presidential offices, wildlife protection agencies, rural residents, pastoralists, and NGO groups within these countries are modeled along with a model of the cheetah-supporting ecosystem. A data set has been collected that consists of political actions by these groups along with cheetah counts by political region. The most practical management plan for this case is computed.
Keywords: ecosystem management, environmental politics, wildlife management, biodiversity, optimal policies.

WIT Transactions on Ecology and the Environment, Vol 108, © 2008 WIT Press
www.witpress.com, ISSN 1743-3541 (on-line)
doi:10.2495/EEIA080241

1 Introduction

Problem: many of the earth's species may be heading for extinction in habitats that are not under the political control of any one country. Those countries that share some of this habitat within their political boundaries, are often developing and have inadequate resources for wildlife conservation – while most of the world's resources for wildlife conservation are expended by developed countries on internal programs. With our current understanding of biological processes, this potential species loss is irreversible. Because of this irreversibility, it can be argued that this problem should be of high priority to all countries. This article gives one way to address this problem.

Because of the above-mentioned developed-country funding bias, lack of conservation funds in habitat-enclosing countries, and lack of political control over these habitats by any one country, conventional conservation programs such as command and control or adaptive management, may not be able to save a managed species from extinction. These considerations have motivated the development here of an approach to ecosystem management that does not assume central control but instead, after building models of both the political processes at work in the habitat-dominating countries and the dynamics of the species-supporting ecosystem, searches for politically feasible ecosystem management plans. Thus, ecosystem management is seen as a two-step procedure: first understand how the political-ecological process works at a mechanistic level and only then begin a search for management plans that require the least change in human behavior patterns in order to effect behavior changes that result in a sequence of actions that leads to the survival of the species being managed. Understanding political processes is seen to be critical because ultimately, the decision to implement ecosystem protection policies is a political one. Management plans that are suggested by examining the output of these ecosystem-only models that ignore political processes may not be supported by the responsible wildlife protection agencies or other affected social groups (hereafter, *groups*) unless the plan addresses the goals of each such group.

This two-step approach to ecosystem management has been implemented in a suite of integrated, web-based analysis tools called an Ecosystem Management Tool (EMT). Use of the EMT allows an analyst to link political processes and goals to ecosystem processes and desired ecosystem outcomes. The EMT's central component is a quantitative, stochastic, and causal model of the ecosystem being managed and the groups involved with this management. This model is called the *political-ecological process simulator* or simply *simulator*. This simulator expresses the group decision making models and the ecosystem model in probabilistic structures known as *influence diagrams* (IDs) (see Pearl [1]). The other components of the EMT are links to data streams, freely-available software for performing all ecosystem management computations and displays, and a web-based archive and delivery system for the first three of these components. The two main uses of the EMT are first to find practical ecosystem management plans, and second to allow anyone with web access to assess for themselves the status of a species being managed with the EMT. This second use is intended to make more accessible to developed countries the status and challenges of managing critical ecosystems in distant, developing countries.

To demonstrate its feasibility, a working EMT for the management of cheetah in a portion of east Africa is developed herein. The portion of east Africa studied is the land enclosed by the political boundaries of Kenya, Tanzania, and Uganda. This ecosystem involves the cheetah, their prey, their physical habitat, and those groups who directly and indirectly manage the cheetah population. Specifically, the simulator models (a) the president, environmental protection agency (EPA), nonpastoralist rural residents (hereafter, *rural residents*), and pastoralists of Kenya, Tanzania and Uganda; (b) a group of nongovernmental organizations (NGOs) that seek to protect biodiversity within these countries; and (c) the cheetah-supporting ecosystem enclosed by these countries.

Once the simulator has been fitted to a political-ecological data set, the simulator can be used to construct the most practical ecosystem management plan. In this article, a practical plan is one that demands the least change in group behavior patterns for a desired improvement in ecosystem health as measured by the ecosystem ID's output nodes. This definition emphasizes political feasibility over a plan's cost. Such a plan is referred to herein as the Most Practical Ecosystem Management Plan (MPEMP).

2 Simulator function and parameter estimation

2.1 Simulator function

The simulator functions by having each group implement an action chosen from a pre-determined repertoire that maximizes their multiobjective (multiple goals) objective function. These actions are in-turn, reacted to by the other groups and may also impact the ecosystem. A temporal sequence of actions taken by groups that affect the ecosystem (the result of playing this sequential game), is called an *ecosystem management plan* (EMP). Such an *actions history* may or may not be the result of a formal, articulated policy for managing the ecosystem. See Haas [2] and [3] for complete descriptions of all group IDs and the ecosystem ID. The model that emerges through the interactions of these IDs is referred to here as an Interacting IDs (IntIDs) model.

2.2 The consistency analysis parameter estimator

IntIDs model parameters are estimated via a procedure referred to here as Consistency Analysis (CA) described next. Let $\mathbf{U}^{(i)}$ be the vector that contains all of the chance nodes that make up the ith ID. IDs 1 through m are models of group perceptions and decision making while the $(m+1)^{\text{th}}$ ID is the model of the ecosystem. Let $\boldsymbol{\beta}^{(Grp)} = (\boldsymbol{\beta}^{(1)\prime}, \ldots, \boldsymbol{\beta}^{(m)\prime})'$ be the stacked vector of group ID parameters wherein $\boldsymbol{\beta}^{(i)}$ parameterizes the ith group ID. Let $\boldsymbol{\beta}^{(Eco)}$ parameterize the ecosystem ID. Finally, let $\boldsymbol{\beta} = (\boldsymbol{\beta}^{(Grp)'}, \boldsymbol{\beta}^{(Eco)'})'$ parameterize the entire IntIDs model. Let $g_S^{(i)}(\boldsymbol{\beta}^{(i)})$ be a *goodness-of-fit* statistic that measures the agreement of the ith ID's probability distribution (referred to here as the $\mathbf{U}^{(i)}|\boldsymbol{\beta}^{(i)}$ *distribution*) and the sample (data set), S. Larger values of $g_S^{(i)}(\boldsymbol{\beta}^{(i)})$ indicate better agreement.

Each parameter in the model is assigned an a-priori value derived from either expert opinion, subject matter theory, or the results of a previous Consistency Analysis. Let $\beta_H^{(j)}$ be such a value assigned to the IntID's j^{th} parameter. Collect all of these *hypothesis parameter values* into the *hypothesis parameter vector*, $\boldsymbol{\beta}_H$. Let $g_H^{(i)}(\boldsymbol{\beta}^{(i)})$ be the agreement between the distribution identified by the values of $\boldsymbol{\beta}_H^{(i)}$ (the $\mathbf{U}^{(i)}|\boldsymbol{\beta}_H^{(i)}$ distribution) and the $\mathbf{U}^{(i)}|\boldsymbol{\beta}^{(i)}$ distribution. As with $g_S^{(i)}(\boldsymbol{\beta}^{(i)})$, larger values of $g_H^{(i)}(\boldsymbol{\beta}^{(i)})$ indicate better agreement. Note that $g_S^{(i)}(\boldsymbol{\beta}^{(i)})$ is the agreement between a sample and a stochastic model, the $\mathbf{U}^{(i)}|\boldsymbol{\beta}^{(i)}$ distribution – while $g_H^{(i)}(\boldsymbol{\beta}^{(i)})$ is the agreement between two stochastic models: the $\mathbf{U}^{(i)}|\boldsymbol{\beta}_H^{(i)}$ distribution and the $\mathbf{U}^{(i)}|\boldsymbol{\beta}^{(i)}$ distribution.

In what follows, the agreement between two distributions will be computed via the *probability density probability function* (PDPF). For a joint event, $\mathbf{u}^{(i)}$ described by an ID, this function is written as $pf_{\mathbf{U}^{(i)}|\boldsymbol{\beta}^{(i)}}(\mathbf{u}^{(i)})$ (see Haas [4]). This function is a generalization of a probability mass function and a probability density function and is necessary because IDs that make up the political-ecological process simulator may contain a mixture of quantitative and qualitative chance nodes.

Let $gmax_S^{(i)}$ be the unconstrained maximum value of $g_S^{(i)}(\boldsymbol{\beta}^{(i)})$ over all $\boldsymbol{\beta}^{(i)}$. Similarly, let $gmax_H^{(i)}$ be the unconstrained maximum value of $g_H^{(i)}(\boldsymbol{\beta}^{(i)})$ over all $\boldsymbol{\beta}^{(i)}$. Up to errors in the approximation of $g_H^{(i)}(\boldsymbol{\beta}^{(i)})$, this value is $g_H^{(i)}(\boldsymbol{\beta}_H^{(i)})$. The Consistency Analysis parameter estimator maximizes the function

$$g_{CA}^{(i)}(\boldsymbol{\beta}^{(i)}) \equiv (1 - c_H)\left[\frac{g_S^{(i)}(\boldsymbol{\beta}^{(i)})}{|gmax_S^{(i)}| + 1}\right] + c_H\left[\frac{g_H^{(i)}(\boldsymbol{\beta}^{(i)})}{|gmax_H^{(i)}| + 1}\right] \tag{1}$$

where $c_H \in (0, 1)$ is the analyst's priority of having the estimated distribution agree with the hypothesis distribution as opposed to agreeing with the empirical (data-derived) distribution. Let $\boldsymbol{\beta}_C^{(i)} \equiv \text{argmax}_{\boldsymbol{\beta}^{(i)}}\{g_{CA}^{(i)}(\boldsymbol{\beta}^{(i)})\}$ be the Consistency Analysis estimate of $\boldsymbol{\beta}^{(i)}$. Hereafter, $\boldsymbol{\beta}_C^{(i)}$ will be referred to as the *consistent* parameter vector for the ith ID. See Haas [4] for further details on Consistency Analysis.

2.3 Agreement functions

2.3.1 Data agreement functions

Call a time series of action-actor-target observations an *actions history* data set. Let $g_S^{(Grp)}(\boldsymbol{\beta})$ be the agreement between the sequence of group actions produced by the IntIDs model and the actions history data set. Let $g_S^{(Eco)}(\boldsymbol{\beta})$ be the agreement between the time series of ecosystem mean values computed by the ecosystem ID and the observed time series of ecosystem values. For the entire IntIDs model, $g_S(\boldsymbol{\beta}) = g_S^{(Grp)}(\boldsymbol{\beta}) + g_S^{(Eco)}(\boldsymbol{\beta})$.

2.3.1.1 Agreement with Actions History Data The Overall Goal Attainment (OGA) node in a group ID represents the group's perceived utility. The action-target combination that maximizes the expected value of this node is the one that the group implements.

Let $\mathbf{out}_i^{(obs)}(t_j)$ be group i's observed output action-target combination at time t_j; $\mathbf{out}_i^{(opt)}(t_j)$ be the action-target combination computed by the group's ID at that time; and M_{ij} be unity if $\mathbf{out}_i^{(opt)}(t_j) = \mathbf{out}_i^{(obs)}(t_j)$ and zero, otherwise. Let $\text{EOGA}_{i,j}^{(obs)}(\boldsymbol{\beta}) = E[OGA^{(i)}|\mathbf{out}_i^{(obs)}(t_j)]$, and $\text{EOGA}_{i,j}^{(opt)}(\boldsymbol{\beta}) = E[OGA^{(i)}|$ $\mathbf{out}_i^{(opt)}(t_j)]$. Then $g_S^{(Grp)}(\boldsymbol{\beta}^{(Grp)}) = \sum_{i=1}^{m} g_S^{(i)}(\boldsymbol{\beta}^{(i)})$ where

$$g_S^{(i)}(\boldsymbol{\beta}) \equiv \sum_{j=1}^{T} M_{i,j} + (1 - M_{i,j})[\max\{\text{EOGA}_{i,j}^{(obs)}(\boldsymbol{\beta}), .001\} - \max\{\text{EOGA}_{i,j}^{(opt)}(\boldsymbol{\beta}), .001\}]. \quad (2)$$

2.3.1.2 Agreement with ecosystem state data Say that a multivariate time series of ecosystem node values has been observed. For example, here, cheetah and herbivore counts are observed over time. Let $\mathbf{u}_{obs}(t)$ be the vector of these values at time t. This vector constitutes a size-one sample on the observable ecosystem ID nodes at t. For such a sample, the negative Hellinger distance is $-|1 - \sqrt{pf_{\mathbf{U}^{(Eco)}|\boldsymbol{\beta}^{(Eco)}}(\mathbf{u}_{obs}(t))}|$ (see Lindsay [5]). Let $g_S^{(Eco)}(\boldsymbol{\beta})$ be the sum of each of these negative Hellinger distances over each combination of region and time point.

2.3.2 Hypothesis agreement function

PDPF values of under an ID's hypothesis distribution, $\mathbf{U}^{(i)}|\boldsymbol{\beta}_H^{(i)}$ and its $\mathbf{U}^{(i)}|\boldsymbol{\beta}^{(i)}$ distribution are approximated by first drawing a size-n sample of design points from a multivariate uniform distribution on the ID's chance nodes: $\mathbf{u}_1, \ldots, \mathbf{u}_n$ – and then estimating $pf_{\mathbf{U}^{(i)}|\boldsymbol{\beta}}^{(i)}(\mathbf{u}_i)$ with an l nearest-neighbor, nonparametric density estimator due to Thompson and Tapia [6] at each of these design points. Using these estimates, the Hellinger distance between $\mathbf{U}^{(i)}|\boldsymbol{\beta}_H^{(i)}$ and $\mathbf{U}^{(i)}|\boldsymbol{\beta}^{(i)}$ is:

$$\hat{\Delta}^{(i)}(\boldsymbol{\beta}^{(i)}, \boldsymbol{\beta}_H^{(i)}) \equiv \left[\sum_{j=1}^{n}\left[\sqrt{\hat{pf}_{\mathbf{U}^{(i)}|\boldsymbol{\beta}_H^{(i)}}(\mathbf{u}_j)} - \sqrt{\hat{pf}_{\mathbf{U}^{(i)}|\boldsymbol{\beta}^{(i)}}(\mathbf{u}_j)}\right]^2\right]^{1/2}. \quad (3)$$

For the collection of group IDs, $g_H^{(Grp)}(\boldsymbol{\beta}^{(Grp)}) = \sum -\hat{\Delta}^{(i)}(\boldsymbol{\beta}^{(i)}, \boldsymbol{\beta}_H^{(i)})$ where summation is over all combinations of time point, group and output node values considered by each group at each time point. For the ecosystem ID, $g_H^{(Eco)}(\boldsymbol{\beta}) = \sum -\hat{\Delta}^{(Eco)}(\boldsymbol{\beta}^{(Eco)}, \boldsymbol{\beta}_H^{(Eco)})$ where summation is over all combinations of region and time point. Then, for the entire IntIDs model the measure of agreement with the hypothesis parameter values is $g_H(\boldsymbol{\beta}) = g_H^{(Grp)}(\boldsymbol{\beta}^{(Grp)}) + g_H^{(Eco)}(\boldsymbol{\beta}^{(Eco)})$.

3 The MPEMP: definition and algorithm

3.1 Definition

Let $\mathbf{Q}(\boldsymbol{\beta})$ be a vector of ecosystem state quantities that are modeled by the simulator's ecosystem ID. For example, $\mathbf{Q}(.)$ could be cheetah count and herbivore count in the year 2058. Assume that a Consistency Analysis has produced a set of consistent parameter values contained in $\boldsymbol{\beta}_C$ and the hypothesis parameter vector has been updated to these values, i.e., $\boldsymbol{\beta}_H = \boldsymbol{\beta}_C$. Using this updated $\boldsymbol{\beta}_H^{(Grp)}$, one way to quantify the concept of a practical ecosystem management plan described in the Introduction is to associate political feasibility with the value of $g_H^{(Grp)}(\boldsymbol{\beta}_{\text{MPEMP}})$ where $\boldsymbol{\beta}_{\text{MPEMP}}$ is a set of group ID parameter values that have been modified away from this updated $\boldsymbol{\beta}_H^{(Grp)}$ so that a desired ecosystem state (expressed as a set of desired values for $\mathbf{Q}(.)$, namely $\mathbf{q}$) is achieved by a sequence of group ID actions over a future time period. The idea is to find a set of minimal changes in group beliefs from those represented by $\boldsymbol{\beta}_H^{(Grp)}$ so that these groups change their behaviors enough to allow the ecosystem to respond in a desired manner. In other words, the MPEMP is the ecosystem management plan that emerges by finding group ID parameter values that result in a desired ecosystem state but that deviate minimally from $\boldsymbol{\beta}_H^{(Grp)}$. Formally, $\boldsymbol{\beta}_{\text{MPEMP}} = \text{argmax}_{\boldsymbol{\beta}^{(Grp)}}\{g_H^{(Grp)}(\boldsymbol{\beta}^{(Grp)})\}$ under the constraint that the ecosystem ID produces output values that are close to the desired ones, $\mathbf{q}$.

3.2 Algorithm to find the MPEMP

First, define an *ecosystem damage utility function*, $f_{\text{ecodam}}(\boldsymbol{\beta}^{(Grp)})$ to be $\sum_{t,i,j} E[OGA_t^{(i)}|\text{ecodam-action}_{i,j}]$ where $OGA_t^{(i)}$ is group i's Overall Goal Attainment node at time t, and ecodam-action$_{i,j}$ is group i's j^{th} ecosystem-damaging action. The sum is over all groups that directly affect the ecosystem, all ecosystem-damaging actions by these groups, and all time points at which such actions are executed during a run of the simulator. Then the MPEMP can be found with the following algorithm.

1. Perform a Consistency Analysis with the current $\boldsymbol{\beta}_H$ and a political-ecological data set to find $\boldsymbol{\beta}_C$. Update $\boldsymbol{\beta}_H$ to this $\boldsymbol{\beta}_C$.
2. Specify $\mathbf{q}$, e.g. 2,000 cheetah and 10,000 herbivores in 2058.
3. Set $k = 1$ and iterate the following three steps:

(a) Compute

$$\boldsymbol{\beta}_k^{(Grp)} = \operatorname{argmin}_{\boldsymbol{\beta}^{(Grp)}} \left\{ \left[\frac{f_{\text{ecodam}}(\boldsymbol{\beta}^{(Grp)})}{f_{\text{ecodam}}(\boldsymbol{\beta}_H^{(Grp)})} \right] + d(\boldsymbol{\beta}) \right\} \tag{4}$$

subject to

$$\frac{|g_H^{(Grp)}(\boldsymbol{\beta}_H^{(Grp)}) - g_H^{(Grp)}(\boldsymbol{\beta}^{(Grp)})|}{g_H^{(Grp)}(\boldsymbol{\beta}_H^{(Grp)})} < 0.1k \tag{5}$$

where $d(\boldsymbol{\beta}) = ||E[\mathbf{Q}(\boldsymbol{\beta})] - \mathbf{q}||/||E[\mathbf{Q}(\boldsymbol{\beta}_H)] - \mathbf{q}||$ and $\boldsymbol{\beta} = (\boldsymbol{\beta}^{(Grp)'}, \boldsymbol{\beta}_H^{(Eco)'})'$.

(b) If $d(\boldsymbol{\beta}_k) = 0$ or $d(\boldsymbol{\beta}_{k-2}) = d(\boldsymbol{\beta}_{k-1}) = d(\boldsymbol{\beta}_k)$, set $\boldsymbol{\beta}_{\text{MPEMP}}$ to $\boldsymbol{\beta}_k^{(Grp)}$ and stop the iteration.

(c) Set $k = k + 1$.

During the optimization, as ecosystem-damaging actions become less attractive to the group, they will not be executed and hence will not contribute to the sum that forms $f_{\text{ecodam}}(\boldsymbol{\beta}^{(Grp)})$. Note that different sequences of group actions can lead to different values in the vector $E[\mathbf{Q}(\boldsymbol{\beta})]$. The utility of such an action needs to be computed under relevant in-combinations, i.e., within the context of the run.

This algorithm finds the MPEMP by sequentially reducing the utility of executing ecosystem-damaging actions under the (gradually-weakened) constraint of staying close to the distribution defined by the updated $\boldsymbol{\beta}_H^{(Grp)}$. Because the smallest changes in values contained in $\boldsymbol{\beta}_H^{(Grp)}$ have been found that achieve the desired ecosystem goals, there are no other group-behavior changes that require smaller changes in group belief systems before such behaviors change enough to achieve the desired ecosystem goals. Hence, the ecosystem management plan that is based on $\boldsymbol{\beta}_{\text{MPEMP}}$ is the most politically feasible.

To implement the MPEMP in the real world, group beliefs that correspond to parameters that have large differences between the updated $\boldsymbol{\beta}_H^{(Grp)}$ and $\boldsymbol{\beta}_{\text{MPEMP}}$ need to be changed in the direction of the $\boldsymbol{\beta}_{\text{MPEMP}}$ value. Methods currently used that attempt to change people's perceptions and values (belief systems) include educational programs, workshops, and advertising.

If the needed degree of belief systems change appears to be beyond available resources, less practical ecosystem management plans can be found by minimizing $f_{\text{ecodam}}(\boldsymbol{\beta}^{(Grp)})$ over parameters that define groups for which beliefs can be realistically changed and leaving the parameters of all other groups at their (updated) $\boldsymbol{\beta}_H^{(Grp)}$ values.

It is possible that the desired ecosystem state values cannot be achieved by the ecosystem ID under any pattern of output actions issued by group IDs. This situation is indicated by $d_i \gg 0$ at the last iteration of the algorithm.

4 Cheetah management example

4.1 Background

Cheetah preservation is a prominent example of the difficulties surrounding the preservation of a large land mammal whose range extends over several countries. The main threats to cheetah preservation are loss of habitat, cub predation by other carnivores, and being shot to control their predation on livestock (Gros [7], Kelly and Durant [8]). In [8], the authors note that juvenile survival is reduced by lion predation inside wildlife reserves because these reserves are not big enough for cheetah to find areas uninhabited by lions. Over-crowding of reserves in Africa is widespread (see O'Connell-Rodwell *et al.* [9]) and cheetah do not compete well for space with other carnivores (Kelly and Durant [8]). Although many cheetah are currently existing on commercial land, this coexistence with man's economic activities may not be a secure long-term solution for cheetah.

One solution would be larger reserves that are free of poachers – possibly enclosed with an electric fence. Such a solution was found to be the most viable for keeping elephants from destroying crops in Namibia (see O'Connell-Rodwell *et al.* [9]). Pelkey *et al.* [10] also conclude that reserves with regular anti-poaching and anti-logging patrols are the most effective strategy for African wildlife and forest conservation.

A large portion of cheetah range is controlled by Kenya, Tanzania, and Uganda (see Kingdon [11]). Currently, the poverty rates in Kenya, Tanzania, and Uganda are 52%, 36%, and 44%, respectively. The adult literacy rates are (90%, 79%) (males, females) for Kenya, (85%, 69%) for Tanzania, and (79%, 59%) for Uganda (World Resources Institute [12]). With close to half of the population living in poverty, many rural Africans in these countries feel that conservation programs put wildlife ahead of their welfare and that large mammals are a threat to their small irrigated patches of ground and their livestock (Gibson [13]). For these reasons, many such individuals are not interested in biodiversity or wildlife conservation.

Gibson [14] finds that the three reasons for poaching are the need for meat, the need for cash from selling animal "trophies," and the protection of livestock. Gibson's analysis suggests that to reduce poaching, policy packages need to be instituted that (a) deliver meat to specific families – not just to the tribal chief, (b) increase the enforcement of laws against the taking of trophies, and (c) improve livestock protection.

4.2 Finding the MPEMP for east African cheetah

Say that a cheetah conservation goal is to have an expected cheetah count of 200 individuals in the Kenyan district of Tsavo 50 years hence, i.e., in the year 2058. Say that only rural resident and pastoralist groups are to have their belief systems modified with all other groups having their parameters held at their updated $\boldsymbol{\beta}_H^{(Grp)}$ values.

Figure 1 shows the Kenya-portion of the political-ecological data set used to find

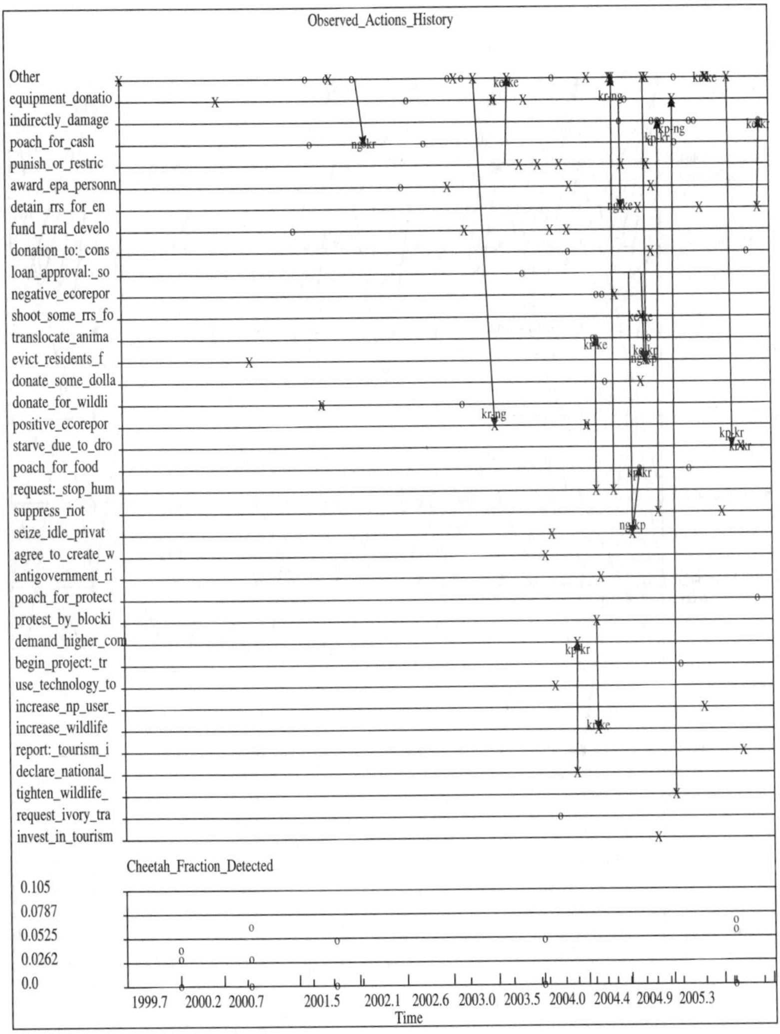

Figure 1: Political-ecological data for Kenya. The ecosystem variable "Cheetah Fraction Detected" is an observable quantity that is derived from the partially unobservable cheetah count, see Haas [4].

the consistent values of the political-ecological process simulator's parameters. In order to show what is predicted to happen if group belief systems remain at their consistent values, Figure 2 shows this consistent model run over the years 2054 through 2058. Figure 3 shows the MPEMP solution over this same time interval. Only the last 5 years are shown because the model quickly enters a repeating pattern of actions and continues this pattern through 2058.

Under the values in the updated $\boldsymbol{\beta}_H^{(Grp)}$, the expected cheetah count in the year 2058 for this district is zero but under $\boldsymbol{\beta}_{\mathrm{MPEMP}}$, it is 176. A comparison of Figures 2

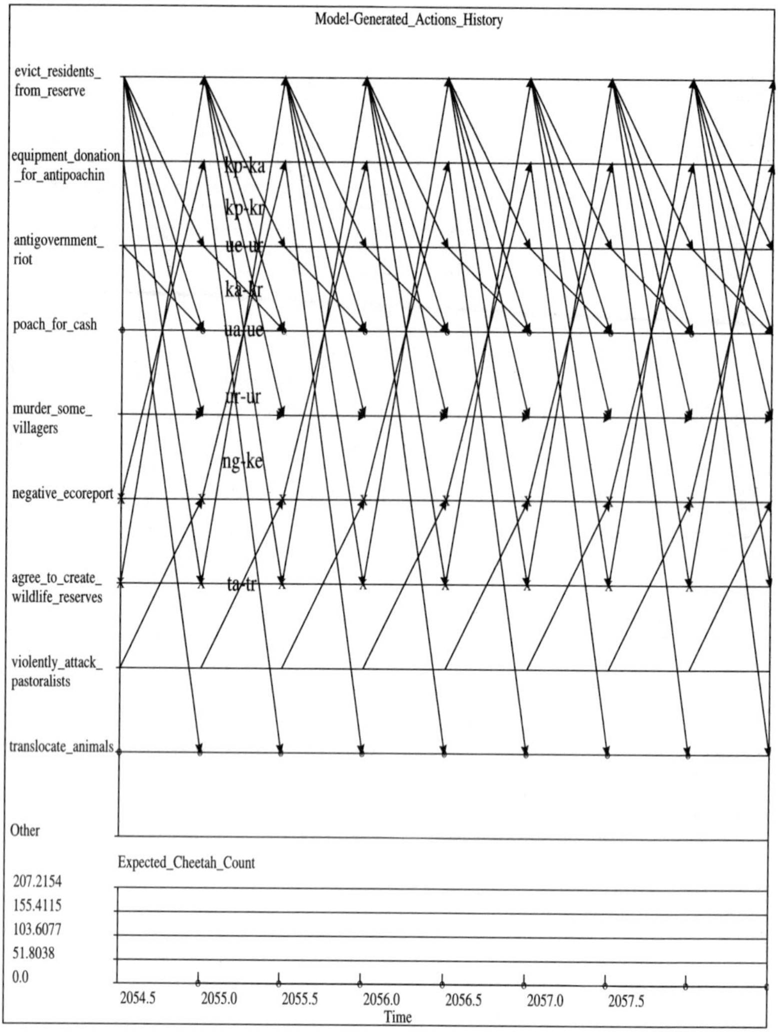

Figure 2: Simulator predictions for 2054–2058 using updated hypothesis parameter values. An arrow's label is located at its center. The first two letters in a label indicates the actor and the second two the target. The first letter in a pair indicates the country (Kenya, Tanzania, or Uganda). The second letter indicates the group type: "r" for rural residents, and "a" for pastoralists. The exception is "ng" for the NGO group.

and 3 reveals that the action: *poach for cash* is not being executed in the MPEMP scenario. This is because the values contained in the updated $\boldsymbol{\beta}_H^{(Grp)}$ that cause this action to be perceived as economically attractive with low risk of arrest are

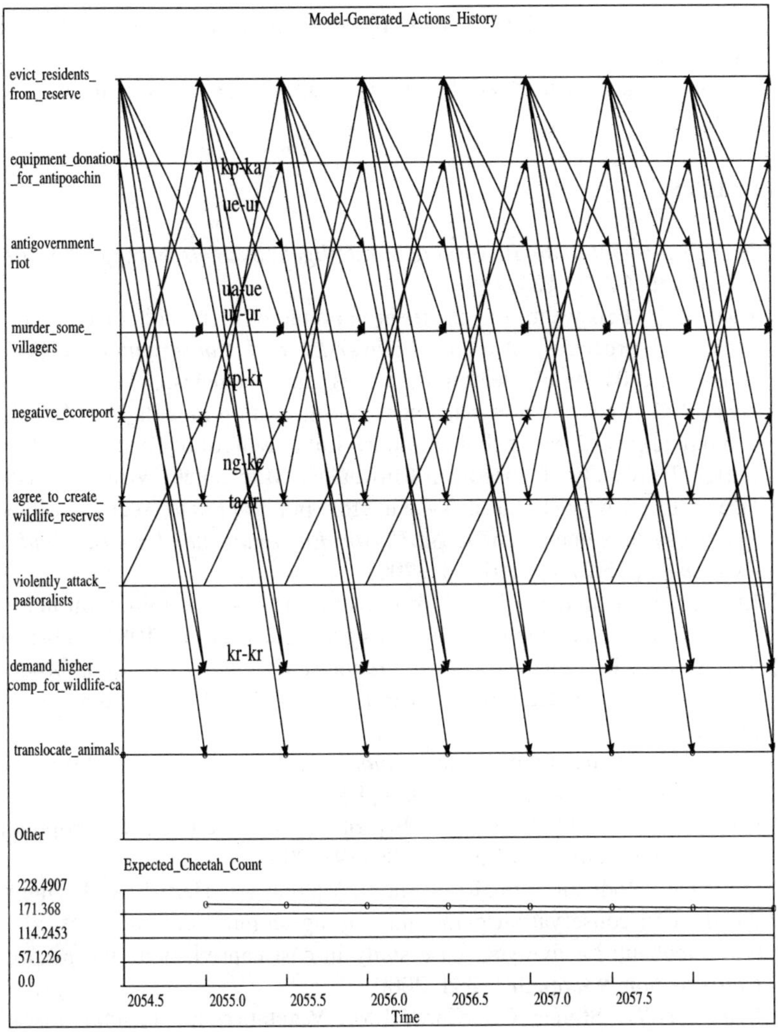

Figure 3: Simulator predictions for 2054–2058 using MPEMP parameter values.

changed to values in $\boldsymbol{\beta}_{\mathrm{MPEMP}}$ that result in the action being perceived as having low economic value and carry a large risk of arrest.

5 Conclusions

This modeling, data fitting, and optimization study shows that it is possible to (a) stochastically model the interactions between political and ecological processes, (b) fit this model to political-ecological data, and (c) use this fitted model to find the most politically feasible ecosystem management plan under one definition of political feasibility.

All software and data used herein is available at www.uwm.edu/~haas/cheetah_emt.

References

[1] Pearl, J., *Probabilistic Reasoning in Intelligent Systems*, Morgan Kaufmann: San Mateo, CA, p. 125, 1988.

[2] Haas, T.C., Ecosystem management via interacting models of political and ecological processes. *Animal Biodiversity and Conservation*, **27(1)**, pp. 231–245, 2004. Available at www.bcn.es/museuciencies.

[3] Haas, T.C., Ecosystem management via interacting models of political and ecological processes. Technical report, Lubar School of Business Administration, University of Wisconsin-Milwaukee, Milwaukee, Wisconsin, 2007.

[4] Haas, T.C., A web-based system for public-private sector collaborative ecosystem management. *Stochastic Environmental Research and Risk Assessment*, **15(2)**, pp. 101–131, 2001.

[5] Lindsay, B., Efficiency versus robustness: the case for minimum hellinger distance and related methods. *Annals of Statistics*, **22**, pp. 1081–1114, 1994.

[6] Thompson, J.R. Tapia, R.A., *Nonparametric Function Estimation, Modeling and Simulation*, Society for Industrial and Applied Mathematics: Philadelphia, p. 179, 1990.

[7] Gros, P.M., Status of the cheetah *acinonyx jubatus* in kenya: a field-interview assessment. *Biological Conservation*, **85**, pp. 137–149, 1998.

[8] Kelly, M.J. Durant, S.M., Viability of the serengeti cheetah population. *Conservation Biology*, **14(3)**, pp. 786–797, 2000.

[9] O'Connell-Rodwell, E.C., Rodwell, T., Matthew, R. Hart, L.A., Living with the modern conservation paradigm: can agricultural communities co-exist with elephants? a five-year case study in east caprivi, namibia. *Biological Conservation*, **93**, pp. 381–391, 2000.

[10] Pelkey, N.W., Stoner, C.J. Caro, T.M., Vegetation in tanzania: assessing long term trends and effects of protection using satellite imagery. *Biological Conservation*, **94**, pp. 297–309, 2000.

[11] Kingdon, J., *East African Mammals: An Atlas of Evolution in Africa*. Academic Press: London, 1977.

[12] World Resources Institute, *Population, Health and Human Well-Being Country Profiles*. World Resources Institute: U.K., 2005. Available at http://earthtrends.wri.org.

[13] Gibson, C.C., *Politicians and Poachers*, Cambridge University Press: Cambridge, U.K., p. 123, 1999.

[14] Gibson, C.C., *Politicians and Poachers*, p. 122. Gibson [13], 1999.

Section 6
Social issues and environmental policies

A study on the corporate social responsibility reports of Greek companies and the use of alternative evaluation methodologies

K. Aravossis & N. Panayiotou
School of Mechanical Engineering,
Sector of Industrial Management & Operational Research,
National Technical University of Athens, Greece

Abstract

Over the last decade there has been an apparent shift from adopting more responsible business practices. To demonstrate that they "care" about people and the environment they operate in, organisations have taken different courses of action. The Corporate Social Responsibility (CSR) reports which are annually published, in addition to the traditional annual financial reports, are considered as one of the vehicles used to demonstrate how caring they have been over the ending financial period and how they intend to continue to be even more proactive in the future. This study examines the adoption of CSR practices in Greece as these are described by companies in CSR reports. The CSR disclosures are further classified and the results are analysed and interpreted in order to draw conclusions and to predict the future of CSR in Greece.

Keywords: CSR, Greece, case study.

1 Introduction: CSR & CSR reporting

Over the last decade there has been an apparent shift from adopting more responsible business practices (Hancock [1]) as a result of regulatory citations, consumer complaints, and special interest group pressures, to proactive research exploring corporate solutions to social problems and incorporating new business practices that will support these issues. Several factors may be contributing to this shift: evidence that socially responsible business practices can actually increase profits; a global marketplace with increased competition and consumer

WIT Transactions on Ecology and the Environment, Vol 108, © 2008 WIT Press
www.witpress.com, ISSN 1743-3541 (on-line)
doi:10.2495/EEIA080251

options; interest in increased worker productivity and retention; and increased visibility and coverage of corporate socially responsible (or irresponsible) activities. Most initiatives relate to altering internal procedures and policies, external reporting of consumer and investor information, making provisions for customer access and privacy, and making decisions regarding suppliers and facility and plant locations.

As with any aspect of a business's performance, corporate responsibility needs to be measured if it is to be understood and managed. However, unlike the more traditional performance criteria such as growth, return on capital, profitability, revenue generation, growth of customer base, etc, corporate responsibility cannot be so easily quantified. Indeed, it has been the lack of absolute measurements that has made more difficult the task of bringing responsibility issues and performance to the attention of both the businesses whose behaviour is being considered and the general public whose awareness of this dimension to business performance is essential for any longterm change in attitudes to take effect. There are signs that the Management of companies believe that CSR can be connected with the achievement of improved financial results (Aravosis *et al.* [2]). Advocates of CSR reports have put forward some perceived benefits, which an organisation may derive from its provision (e.g. Crowther [3]). Typical examples include: increased customer loyalty, more supportive communities, the recruitment and retention of more talented employees, improved quality and productivity and the avoidance of potential reputational risks which may arise from environmental incidents (Idowu and Towler [4]). However, Cooper [5] noted that the practical experience of early adopters of CSR reports was mixed. For example, instead of enhancing companies' reputation, CSR reports attracted adverse comments by drawing attention to divergences between the values espoused by the company and its actual behavior.

The remaining of the paper analyses the CSR initiatives in Greek companies.

2 Research methodology

We contacted all companies involved with CSR activities that publish CSR reports. In order to identify such companies, we contacted the most important CSR organizations: Hellenic Network for Corporate Social Responsibility (CSR Hellas), Centre for Sustainability and Excellence (CSE) and Eurocharity. Moreover, the CSR practices of companies listed in the Athens Stock Exchange were analyzed. Also, we contacted selected company executives that had participated in CSR–related conferences in Greece, in order to obtain in depth understanding of their CSR initiatives and obtain information concerning their latest CSR reports. As a result, we came in contact with executives from three international companies. The interviews we had, worked as pilots and helped us to better understand the CSR reality in Greece.

148 companies with registered offices in different parts of Greece were contacted by telephone and/or e-mail in order to obtain information concerning their CSR practices 15 of the contacted companies were members of more than one

CSR organizations. Our contacts permitted us to identify the companies disclosing CSR information by different methods. Interestingly, some of them offered to send us their organisations' annual reports or suggested putting us through to a more senior individual in the organisation who might be able to help us.

In order to understand how the companies contribute positively to societal lives, the research analysis was designed to highlight the companies' disclosures in industries with insignificant impact on the environment. We received information from 81 companies (about 55% of the organisations contacted).

Figure 1 summarises the research approach followed, highlighting its distinctive phases and its related activities.

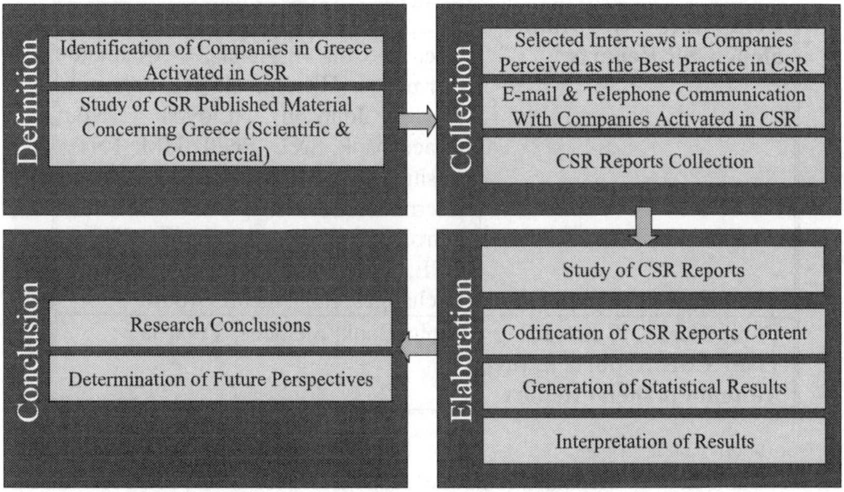

Figure 1: CSR research approach.

3 Research findings

From the information received, it was found that 28 organisations report CSR results in Greece. These organisations have adopted two different ways of reporting CSR activities. Some organisations issue a stand-alone CSR report, whilst others devote a section in their corporate annual report for CSR activities.

Table 1 summarises the identified methods used by companies for CSR reporting in Greece.

Another classification of the contacted companies was based on the Organisational Unit, which is responsible for CSR issues. Three different practices were identified in Greece:

- A dedicated CSR Department is responsible for Corporate Social Responsibility: Such a practice shows the importance placed in the CSR processes by the organisation. In the case of Greece, 17 companies have a separated CSR department.
- CSR processes are carried out by the Public Relations Department: Such a practice implies that CSR is seen as an adjunct of Public Relations, a

function of a company's external relationships, a peripheral activity, not something that needs to be embedded across the organisation horizontally and vertically.

- CSR processes are carried out by another executive: Similarly with the above practice, this one may imply that CSR is perceived as a peripheral activity, with small importance, limited to participating in social and economic regeneration initiatives and supporting the work of charities and voluntary bodies.

Table 1: Methods of CSR reporting of companies operating in Greece.

Companies with Stand-alone CSR Reports	BP, Club Hotel Casino Loutraki, Coca Cola Hellas, Cosmote Germanos, Hitachi, HSBC, Johnson & Johnson, Lloyds TSB, Piraeusbank, S&B, Shell, TIM, TNT, Toshiba, Vodafone, Titan, Athens International Airport, Hellenic Petroleum, Emporiki Bank, Opap, OTE, Friends Provident, Hellenic Exchanges, Italcementi Group
Companies Embedding Their CSR Reports in the Annual Financial Report	Alpha Bank, Siemens, Vivartia

Table 2: Organisational unit responsible for CSR processes.

Companies With an Independent CSR Organisational Unit	**Companies That Make Responsible for CSR the person in charge of public relations**	**Companies that make responsible for CSR another executive**
Alpha Bank, Cosmote, BP, OPAP, Vodafone, Hitachi, Piraeusbank, Toshiba, Coca-Cola Hellas, TNT, Titan, Shell, Johnson&Johnson, Lloyds TSB, Germanos, HSBC, Friends Provident	Club Hotel Casino Loutraki, S&B, TIM, Vivartia, Italcementi Group, OTE, Hellenic-Petroleum, Hellenic Exchanges	Siemens, Emporiki bank, Athens International Airport

Table 2 presents the companies included in the three above categories.

The great majority of the companies that issue a CSR report are international companies. This first approach evinces that the companies that are more active in the field of CSR are companies with a wider range of activation. It follows that international companies are more influenced by new trends and also they should demonstrate a good sociable and environmental behavior in order to be sustainable.

Another classification criterion of the companies was their status of ownership. Two different categories were recognised based on this criterion: private organizations and public organizations. The first category includes all companies where the majority of shares belong to private individuals. The second includes all companies where the majority of shares belong to the state. Table 3 displays the fact that the majority of organizations are private, which in combination with the above-mentioned leads to the conclusion that the companies that are innovative in CSR are those that belong to private individuals and their activation surpasses the borders of Greece. It is a little bit suspicious that there are only three public organizations among those that issue CSR reports. Traditionally, public organizations intent to offer to the public more compared to the private ones. The fact that CSR is not so popular among public organizations may have its roots to bureaucracy, which leads to slow adoption of new ideas.

Table 3: Classification of companies according to the status of ownership.

Private Organisations	Vodafone, Titan, Piraeusbank, Vivartia, Hitachi , TNT, Toshiba , BP, HSBC, Friends Provident, Italcementi Group, Shell, S&B, Siemens, Coca-Cola 3E, Club Hotel Casino Loutraki, Lloyds TSB, TIM Hellas, Germanos, Emporiki Bank, OTE, Johnson & Johnson, Cosmote, Alpha Bank, Hellenic Exchanges
Organisations With Governmental Participation	Athens International Airport, Hellenic-Petroleum, OPAP

Companies were further classified based on their sector of operation. Companies that publish CSR reports in Greece belong to eleven different sectors: telecommunication, financial services, logistics and transportation, industry, cement industry, mining and metals, refinery, distillery, dairies, leisure and entertainment and public agencies. Most companies that issue a CSR report belong to the financial services sector. The telecommunications and manufacturing sectors follow.

Our research also displays that the companies that issue CSR reports are big companies concerning the number of employees they occupy, with a medium size price of 73,455 employees.

Table 4 summarizes the results of the study concerning the CSR practice in Greece as this is revealed by the contents of Corporate Social Responsibility Reports of the analyzed companies.

Table 4: Summary of CSR findings in Greece.

Issues	Findings
Number of companies active in CSR issues in Greece/ Publishing a CSR Report in Greece	81/28
Number of companies in Greece/ Number of companies listed in the Athens Stock Exchange	More than 800,000/ 310
Most frequent method of CSR reporting	Stand-alone CSR Reports (25/28)
Size of companies	79,031 employees
Scope of operation	International (24/28) National (4/28)
Regime of ownership	Private Sector (25/28) Extended Public Sector (3/28)
Number of companies with CSR reports listed in Athens Stock Exchange or in other Stock Exchange	28/28
Dominating Industrial Sectors	Financial, Communications, Petroleum – Refinery, Manufacturing
Recognised CSR categories	Economy, Internal Business Processes, Learning, Environmental Impact, Human Resources , Society Marketplace, Health & Safety
Dominating Organisational Unit responsible for CSR	CSR Department Department of Public Relations
Number of different performance measures used	303

4 Conclusions

Despite the diversity of companies and CSR material selected for investigation, it does seem possible to draw carefully some tentative conclusions about CSR practice in Greece.

First, the penetration of CSR in Greece is definitely small so far, a fact proved by the small number of published CSR reports and the small number of companies participating in organisations activated in CSR issues. However, there are indications in the Greek market (identified by personal interviews with CSR experts in Greece) that it will increase in the following years.

Second, CSR does not appear to be a systematic activity. To the extent that it is not covered by regulations, social disclosure seems to wax and wane in popularity, in the subjects to which it gives attention and in terms of the organizations which provide such disclosure.

A third conclusion of the study in Greece concerning CSR is that its adoption by the companies does appear to be related to certain 'demographic' characteristics. In particular:

- The size of the company plays an important role in its CSR orientation. It was found that large international corporations are the majority of the organisations in Greece that publish CSR reports and disclose CSR-related information in a systematic way. The parameter of international operation is considered as very important revealing that there could be a significant correlation of CSR orientation and the origin of ultimate ownership of the company. This last issues demands further research in the future.
- There is some evidence that the sector of operation affects the degree of CSR adoption. Companies operating in Financial Services, Telecommunications and the Petroleum Industry show a much higher penetration rate compared to all the other sectors. It would be very interesting to further examine the reasons for the behaviour of these sectors.

The classification of performance measures used by companies was a very difficult process of our study. The number of the identified performance measures proves the existing degree of complexity and highlights the luck of:

- A commonly understood definition (within and across companies).
- A common set of benchmarks to measure the attainment of corporate social responsibility.
- Formal established processes in place to achieve these benchmarks.
- A system of internal auditing of the CSR processes.
- A system of external verification by accredited bodies.

It has to be stated, however, that the fact that companies are beginning to accept they have to account in some form for their wider impact on society is a significant step. The methodologies behind these social reports may be poor and their terms of reference self-serving for the company, but the commitment is an important one. As the concepts of corporate social responsibility become more clear and specific, companies will become more sophisticated in their social reporting.

Finally, CSR does not appear to be related to profitability in a documented quantitative way. This finding is not surprising, taking into account the abstract framework of performance measures.

In sum, our empirical study suggests that CSR in Greece is in its infancy phase. Its adopters are mainly large international companies; however, there are aspects of developments in disclosure practices. Some of the companies seem to believe in CSR concepts, others are just following the practices of the best and some appear to use the report as a public relations exercise. CSR reporting needs further improvement as the information it contains needs to be standardized. Taking into account that there is an increasing number of companies in Greece that are truly concerned about the environment, resources, quality of life and social matters but are not yet aware of CSR principles, it is believed that the example of CSR reporting by the large and successful companies as well as an official encouragement of social responsibility by the Greek Government in the future will lead the way towards a conscious responsible behaviour of companies operating in Greece.

References

[1] Hancock, J., Investing in Corporate Social Responsibility - A Guide to Best Practice, Business Planning & the UK's Leading Companies. London: Kogan Page Ltd, 2005.

[2] Aravossis K., Panayiotou N. & Tsousi K., A Proposed Methodological Framework for the Evaluation of Corporate Social Responsibility In: K. Aravossis, C.A. Brebbia, E. Kakaras & A.G. Kungolos, eds. Environmental Economics and Investment Assessment, Southampton UK, WIT Press, pp. 87–95, 2006.

[3] Crowther, D., Corporate social reporting: genuine action or window dressing?, In: D. Crowther, & L. Rayman-Bacchus, eds. Perspectives on Corporate Social Responsibility, Ashgate, Aldershot, pp. 140–60, 2003.

[4] Idowu S.O. & Towler B.A., *A Comparative Study of the Contents of Corporate Social Responsibility Reports of UK Companies*, Management of Environmental Quality: An International Journal, Vol. 15 No. 4, pp. 420–437, 2004.

[5] Cooper, B., Corporate social responsibility: the Holy Grail?, Chartered Secretary, July, pp. 12–16, 2003.

An empirical study of what institutions of higher education in the UK consider to be their corporate social responsibility

S. O. Idowu
London Metropolitan University, UK

Abstract

The field of corporate social responsibility (CSR) is relatively new when compared with other fields. It was in the 1980s that corporate entities around the world started to generate increasing interest in CSR, as we currently know it. However, a search of the literature has revealed that researchers in the field have tended to concentrate more efforts on the CSR of profit seeking corporate entities. Unfortunately this does not appear to be the case when it comes to the CSR of not-for-profit (NFP) corporate entities such as educational establishments (schools, colleges and universities), hospitals, the police, the armed forces, the fire services and other social entities that play equally important roles in modern economies. Do these NFP organisations consider that they have a role to play in CSR as is perceived by corporate stakeholders in the 21st century? If they do, then do they play these roles as effectively as they should? How do these entity stakeholders perceive the contributions they make to society's well-being? These and other pertinent questions are what this research study seeks to find answers to.

The study looks at what the UK's institutions of higher education consider to be their corporate social responsibilities and how they have absorbed these responsibilities into what they do in order to discharge these responsibilities to local, national and international communities. The research shows that most institutions of higher education in the UK are conscious that they owe some responsibility to all their stakeholders and are striving to demonstrate this awareness in various ways.

Keywords: social responsibility, not-for-profit organisations, higher education institutions, environment, stakeholders, universities, UK, sustainability.

WIT Transactions on Ecology and the Environment, Vol 108, © 2008 WIT Press
www.witpress.com, ISSN 1743-3541 (on-line)
doi:10.2495/EEIA080261

1 Introduction

Corporate entities of whatever shape or form (profit motive or not-for-profit) aspire to be successful in whatever they do. Success can come about through several avenues. In recent times, corporate entities around the world have come to realise that success can be achieved when they are perceived by their stakeholders as being socially responsible; these stakeholders tend to warm to what the entities do or stand for; which consequently makes a big difference in terms of whether or not they achieve their strategic objectives. A socially responsible corporate entity takes cognisance of the impact of its actions on its communities, its stakeholders and the environment when formulating its corporate objectives and in its decision making process. It strives at all times to either minimise or totally remove the adverse effects of its activities on the environment, employees, business contacts group, suppliers of funds and credits, governments and other affected members of society. Corporate entities around the world now consider that being socially responsible is very 'trendy' indeed!

The activities of several non-governmental organisations (NGOs) (also known as pressure groups) can be argued to have played a catalytic role in the emergence of the field of corporate social responsibility. These NGOs try to police different aspects of corporate and governmental behaviours in both the developed and less developed countries of the world in order to ensure that these behaviours do not fall short of acceptable standards. International organisations such as Green Peace, Amnesty International, Friends of the Earth, World Wildlife Fund are a few of such organisations. In addition, business associations such as the World Council on Sustainable Development, Business in the Community (BitC), CSR Europe and several others are propagating various CSR related activities and events to increase corporate and societal awareness of what the field entails and how beneficial its inculcation into business operations would be on the environment, their stakeholders and the world at large. The United Nations and many of its arms have been very active in the area of Sustainable Development; over the last few years several United Nations' sponsored related conferences have been held, the Bali Climate Change Conference in December 2007 is a recent example. Several governments across the world are also propagating CSR. The present UK government continues to portray itself as an enthusiastic advocate of CSR.

Universities across the world because of the societal leadership role they have through their research and particularly their teaching are in a unique position to make a difference in the practice and acceptance of CSR by tomorrow's leaders. This therefore explains why a study on universities involvements with CSR will be of interest to us all.

The higher education sector in the UK has been experiencing a transformation phase of late. Institutions of higher education are gradually becoming more *self-autonomous*, students are now perceived as *clients*, the word *'competition'* was of no significance in the industry before now, there was no distinction between local and overseas students *vis-à-vis* the fees they paid until the 1980s and *profit* or *surplus* was not one of the variables used previously to measure success or performance; as long as the objectives of the institution concerned were achieved in terms of efficiency, effectiveness and value for money. All these are changing in UK's higher education institutions and it appears that nothing can be taken for granted anymore! The industry will become even more competitive and self-regulated as time goes by.

The system of higher education has now been internationalised according to the International Finance Corporation [18], which argues that in 2003 about 2% of the world's university student population of 100m were studying outside their home countries. It also notes that there has been an annual growth of 7% in the market since the late 1990s with a current annual fee income of $30 billion. A fee income of this magnitude signifies the enormity of the market. If institutions of higher education in the UK wish to continue to attract foreign students who are prepared to pay higher discriminatory fees in order to enable them to take a reasonable portion of that total annual fee income, in addition to providing good quality services (which they are noted for) they must be seen to be socially responsible by this stakeholder group (*clients*) and other stakeholder groups related to them e.g. governments of their home countries, prospective employers and loan providers such as the International Education Finance Corporation (an arm of the World Bank which provides loans to international students).

This paper is constructed as follows: first, it reviews the literature on the corporate social responsibility of profit seeking corporate entities and the few available ones on not-for-profit corporate entities with a description of some of the previous studies undertaken in the area. The empirical work undertaken for the study is presented next, including details of the method used and its findings. Finally, the implications of the paper are considered with the author's concluding remarks.

2 Reviewing the literature on CSR as it applies to corporate entities including universities

The concept of corporate social responsibility is not a totally new one as noted by several academic researchers in the literature; the British Institute of Management noted that the use of CSR has been on the corporate scene in the UK for as far back as 1947, Crowther [5], who argues that CSR has been in existence in Britain since the Industrial Revolution of the 18th century, Maltby [22], who argues that it has been practiced proactively by a number of British manufacturing companies; especially Sheffield steelmakers during the beginning of the 20th century and Norris and O'Dwyer [24] who summed it all up when they argued that 'the concept has received much attention in the past but this has tended to wax and wane; what we are now witnessing is only a resurgence of interest in the field of corporate social responsibility'.

Academic researchers, in an attempt to understand the factors that have helped to heighten recent interests in the field of corporate social responsibility as we presently understand it; have postulated a number of theories which they have used to support their arguments and formalise the results of their studies. These are *decision usefulness* theory (which argues that investors find the social information disclosed by corporate entities helpful in their decision making exercise) see for example Spicer [27], Buzby and Falk [2], Belkaoui [1], Mahapatra [21], *agency* theory (which views the relationship that subsists between the managers and owners of a corporate entity as that of agents and principals) see for example Jensen and Meckling [19], *stakeholder* theory (which assumes that in order for an entity to generate sustainable wealth over a long period of time, good relationships must exist between that entity and its critical stakeholders) see for example Carroll [3], Freeman [8], Clarkson [4] (who distinguishes between two classes of stakeholders namely; primary and secondary stakeholders), *sustainable development* theory (which argues that the future of mankind lies in his ability to build sustainable business enterprises and an economic reality which connects industry, society and the environment) see Hart [12], Senge and Carstedt [26] and Ricart *et al* [25], *social and political* theory (which describes society as operating under 'a series of social contracts between members of society and society itself) Gray *et al* [11] and finally *legitimacy* theory (which postulates that 'the actions taken by an entity are assumed to be desirable, proper or appropriate within some socially constructed system of norms, values, beliefs and definitions') Lindblom [20], Suchman [28]) and Gray *et al* [10].

Vogl [30] takes a slightly different approach to the theories mentioned above. He argues that four factors can be recognised as contributing to the recent trends in corporate entities around the world embarking on socially responsible behaviours. These factors he argues; are tightening regulatory pressures, changing demographics, pressure from non-governmental organisations and the increased necessity for greater transparency.

Prior to the 1970s, the maximisation of the shareholders' wealth was assumed to be the singular objective of the firm. Thus Friedman [9] argues that managers are expected to conduct the business of the enterprise in accordance with the owners' or shareholders' desire, which generally will be to make as much money as possible whilst conforming to the basic rules of society which are embodied in local laws and customs. However, Elkington [6] argues that this cannot possibly be the case as shareholders only belong to one of the many stakeholder groups of a business entity. The objectives are in actual fact three-fold argues Elkington, namely to create economic value (which agrees with Friedman's argument), ecological and social values for all concerned (that is all affected stakeholders). It should be noted that there is a twenty-seven year gap between their times. Friedman's argument was not wrong. It was only based on what were then the accepted standards. Elkington's argument was equally not incorrect, by 1997 society had become more developed, more sophisticated and better informed, corporate entities too had recognised that they operate in a multi-stakeholder world and that they must be seen to take on board the needs of all their stakeholders. They can no longer continue to meet the needs of just one stakeholder group at the expense of the other stakeholder groups; a balance needs to be struck in meeting the needs of them all.

The issue of sustainability (which is the bigger umbrella under which CSR falls) in higher education has become a big debate and consequently a research based area for academics in various institutions of higher education around the globe see for example van Weenen [29] whose research looks at the experiences of different universities around the world during the process of integrating sustainable development in their activities including the university of Amsterdam, The Netherlands, Noeke [23] whose research in Germany focuses on the establishment of an environmental management system at the University of Paderborn, Wright [32] whose research in Canada focuses on a set of major national and international frameworks for sustainability in higher education through the use of declarations, Wals and Jickling [31] (who were based in The Netherlands and Canada respectively) their research explores both the overarching goals and process of higher education from an emancipatory view and with regard to sustainability, Fien [7] whose research in Australia explores issues related to the choice of goals and approaches for advancing sustainability in higher education and Holt [15] whose research examines the values, actions and attitudes of a group of students as they enter and leave a business school at Middlesex university UK. This has come about as a result of different factors. Firstly, societies are becoming increasingly conscious that man's natural resources nature has will not last forever; in fact some species are gradually becoming extinct at an alarming rate (Wals and Jickling [31]) and if nothing is done at the rate some of these resources are being used or wasted they will become too scarce when the next generation occupy this planet. Secondly, international organisations for example the United Nations and its various arms are organising conferences and conventions on Sustainable Development and Climate Change to draw the attention of world leaders in the fields of politics, education, business and others (who have the power to influence behaviours); that actions need to be taken urgently to alleviate future problems in this area; see for example *Agenda 21* in Rio of 1992, the *Kyoto Protocol* of December 1997, Bali Climate Change Convention 2007 and similar actions which have been taken. Finally, it is apparent that educators in higher education institutions are in fortunate and privileged position because they have the opportunity and power to influence the thoughts, future actions and behaviours of tomorrow's managers and leaders today! It therefore makes sense to start the 'indoctrination' of these future leaders about these issues right from these institutions of learning.

3 Corporate social responsibility and sustainable development

CSR is an all embracing word that covers a wide variety of activities. Some authors have described the field as being about corporate entities' ability to 'in addition to making a profit, also helping to solve some social problems regardless of whether or not they have been responsible for creating those problems in the first place; even if there is potential profits or gains either in the short-run or long-run' Holmes [14]. The field is about what corporate entities are doing to contribute positively to what is going on in the world around them. It is now a common practice for organisations in certain industries and in the more industrialised parts of the world to install environmentally friendly machinery, use recyclable raw materials, rehabilitate sites which may have been damaged by their previous actions, treat employees equally regardless of sex, race, religion, disability etc, respect the conventions on human rights, disassociate themselves from suppliers of child labour products, be engaged in sustainable development, make donations for charitable purposes and other socially responsible actions which modern corporations embark on in order to demonstrate that being responsible is as important to them as doing well Idowu and Papasolomou [17].

Sustainable development is an important aspect of CSR and corporate entities (regardless of whether profit oriented or not profit oriented) which aspire to be perceived by their stakeholders as being socially responsible must be interested in sustainability and sustainable development. In an attempt to encourage institutions of higher education to be involved in sustainable development, a new initiative was established in the summer of 2000 funded by the UK Higher Education Funding Councils in England, Scotland, Wales and Northern Ireland and led by the Forum for the Future called the Higher Education Partnership for Sustainability (HEPS). HEPS is made up of 18 universities and colleges in the UK, twelve were drawn from England, four from Scotland and one each from Wales and Northern Ireland. HEPS' aim was '*to establish a pioneering partnership group of Higher Education Institutions (HEIs) that are seen to be achieving their strategic objectives through positive engagement with the sustainable development agenda, and to generate the transferable tools, guidance and inspiration that will encourage the rest of the sector to do likewise*'.

HEPS now believes that the 18 universities and colleges they have been working with are now delivering excellent education in a way that boosts sustainable development because all are now developing green buildings, cutting emissions and improving the curriculum. In HEPS' view, universities and colleges play three roles in society, as:

- Institutions that form and inform leaders and decision-makers of today and tomorrow through teaching and research agendas.
- Managers of major businesses where prudent use of resources not only saves money but safeguards reputations.
- Important bodies in the local communities and regional development- as employer, purchaser, service user and provider.

The above three roles emanate from the speech delivered by D. L. Johnston, the then Principal and Vice-Chancellor of McGill University, Canada and Member of the International Association of Universities (IAU) Administrative Board of November 1993 at the IAU 9th Round Table, Kyoto, Japan.

In order to ensure commitment to active engagement in the partnership, agreement was reached at vice-chancellor/principal level. The work of HEPS is delivered through three types of activities:

- Individual work programmes tailored to the institution's priorities.
- Partnership wide capacity building activities covering areas of interest to all partners (e.g. purchasing, travel planning, finance, resource management and communicating for sustainability).

- ➢ Sustainability reporting: developing a framework and process for tracking progress and communicating an institution's contribution towards sustainable development.

Source: Accounting for Sustainability (2003) p. 7.

4 The research study

The study seeks to contribute to knowledge in the field of corporate social responsibility in relation to a group of not-for-profit social entities using higher education institutions in the UK. Institutions of higher education are classified in the UK as not-for-profit or public sector corporate entities. It was felt that the results of such research would enrich people's understanding of what these institutions are doing about social responsibility and how they go about discharging their responsibilities to societies both far and near.

To gather the information needed for the study, letters were sent to one hundred and twenty-one (121) Vice Chancellors/Principals/Chief Executives of all universities and colleges of higher education in the UK – 92 in England, 14 in Scotland, 2 in Northern Ireland and 13 in Wales. The letter briefly explained what the study was about and asked the Vice Chancellor or Principal or Chief Executive to 'explain in as much detail as possible what he/she considers to be the social responsibilities of his/her university/college and how the university's management goes about discharging these responsibilities'. A conscious decision was made not to send reminders to non-responding VCs for two reasons. To ensure that no one was coerced into participating in the study since these individuals are generally recognised as extremely busy people and secondly to establish whether being socially responsible is now part of normal day to day practice of these institutions of higher education.

Thirty-three (33) replies were received from 33 universities and colleges. Replies from these institutions have been analysed and grouped according to their themes and commonalities. A conscious decision was made not to name or ascribe any piece of information to any university. This was to ensure anonymity and to ensure that non responding universities were not disadvantaged in any way as a result of the study. Out of the 33 replies received, 10 universities told stated that they were unwilling to provide the required information, giving different reasons for their decision.

Twenty-three higher education institutions from the four UK countries have participated in the research and their names are: Birmingham, Brighton, Bradford, Buckingham, Derby, Hertfordshire, Hull, London Metropolitan, London South Bank, Loughborough, Luton, King's College London, Imperial College London, Manchester, Nottingham, Roehampton, Royal College of Music London, Wolverhampton, Queen's Belfast, Cardiff, Aberdeen, Glasgow and Glasgow Caledonia.

The list above indicates the following response rates from HE institutions in the four UK countries England 19.57%, Northern Ireland 50%, Scotland 21.43% and Wales 7.69%. It also indicates that 61% of these responding universities are pre-1992 universities and 39% are post-1992 universities.

The aspects that these institutions consider to be their corporate social responsibility have been grouped under different headings; some of these aspects are common to them whilst a few are unique to certain universities. The following are the general headings:

- Widening Participation
- Developing and communicating performance on sustainable developments
- Contributing both to national and international systems of university education
- Managing the economic, social and environmental impacts of their activities
- Take into account the interests of all stakeholders and act as good citizens
- Joining Business in the Community (BitC) as an initial step
- Responding to social needs in terms of Education and Innovation
- Engagement with Corporate bodies through Staff
- Providing a more effective Community Service
- Challenging, Inspiring and Supporting Students to Grow
- Sustaining and adding value to country's culture, economy and the natural environment
- To manage and govern itself with responsibility and sensitivity

5 The implications of the findings from the research

The findings of this research study have clearly demonstrated that some institutions of higher education in the UK are still not explicit of what their responsibilities to society are. Some universities are still a few years behind others when it comes to demonstrating their social responsibilities to stakeholders. A document in place on the organisation's policies or frameworks on CSR would have sufficed. We received excuses such as –'we are too busy' 'we receive too many requests similar to this' 'you are not a big organisation' etc, perhaps they do not have any policy or documents to give callers on their social responsibilities. It is hoped that this study would serve as a wake-up call to any institution that falls within this category. Being socially responsible is now no longer an option, but a moral obligation, which must be fulfilled by all responsible corporate entities. CSR in higher education institutions is not about teaching and researching, society has accepted that these institutions are very proficient at doing these. It is about taking cognisance of the impact of their operations on their stakeholders, the environment and the world at large. Except they can demonstrate beyond reasonable doubts that being socially responsible is equally important to them and is aware of what this entails; they have a very high mountain to climb. The industry they operate in is rapidly changing; they must rise up to all the challenges that accompany the inevitable change. They can do it, because they are really doing a good job in their core business, all they need to do is find the time to do what is required.

6 Conclusion

This study confirms that many UK institutions of higher education are taking the issue of corporate social responsibility and sustainable developments seriously. Many, as far as one can see are socially responsible; providing non-sensitive information to members of the public who need this for whatever purposes actually demonstrates a very high degree of responsibility. Not many of these institutions fall under this category. The industry is changing. They now have to deal with a host of stakeholders, *clients* and *competitors* included. Like their counterparts in the profit seeking sector; which have since realised that in order to compete effectively in modern markets; to out-perform their counterparts in their industry they need to have some competitive edge, CSR could perhaps provide this for them. They can no longer continue to just conform to the basic rules of society; in fact many of those basic rules are changing too and would hopefully continue to change. They must go well beyond meeting society's basic expectations in other words, conduct themselves in a socially desirable manner. Unless, these institutions re-define their corporate strategies many of them are unlikely to survive and prosper in the 21st century as the industry becomes more competitive.

Many of these institutions have realised that they cannot only be involved with impacting educational knowledge and giving paper qualifications – their social responsibility has extended well beyond that boundary, they must also be seen to be actively involved in helping to solve social problems which fall within their areas of confine. Being socially responsible, appears to be the way forward not just for profit seeking corporate entities but also for those entities in the not for profit sector of the economy.

References

[1] Belkaoui, A. (1980), 'The impact of socio-economic accounting statements on the investment decision: an empirical study', *Accounting, Organisations and Society* Vol. 5 No. 3 pp. 263–283.

[2] Buzby, S. L. and Falk, H., (1979), 'Demand for social responsibility information by university investors', *The accounting Review* Vol. 54 No. 1 pp. 23–37.

[3] Carroll, A. B. (1989), Business and Society: Ethics and Stakeholder Management, South-Western, Cincinnati, OH.

[4] Clarkson, M. B. E. (1995), 'A stakeholder framework for analysing and evaluating corporate social performance' *Academy of Management Review,* Vol. 20 pp. 92–117.

[5] Crowther, D. (2003), 'Corporate social reporting: genuine action or window dressing' In Crowther, D. and Rayman-Bacchus, L. (Eds.), *Perspectives on Corporate Social Responsibility,* Ashgate, Aldershot, pp.140–160.

[6] Elkington, J. (1997), Cannibals with Forks: The Triple Bottom Line of 21st Century Business, Capstone, Oxford.

[7] Fien, J. (2002), 'Advancing sustainability in higher education: Issues and opportunities for research' *International Journal of Sustainability in Higher Education*, Vol. 3 No. 3, pp. 243–253.

[8] Freeman, R. E. (1984), Strategic Management: A Stakeholder, Pitman Publishing, Boston, Mass.

[9] Friedman, M. (1970), 'The social responsibility of business is to increase its profits' *The New York Times Magazine,* September 13.

[10] Gray, R., Walters, D., Bebbington, J, and Thomson, I. (1995), The greening of enterprise: an exploration of the role of environmental accounting and environmental accountants in organisational change, *Critical Perspectives on Accounting*, Vol. 6 No. 3 pp 211–239.

[11] Gray, R., Owen, D. and Adams, C. (1996), 'Accounting and Accountability, Changes and Challenges in Corporate Social and Environmental Reporting', Prentice-Hall, Harlow.

[12] Hart, S. L., (1997), 'Beyond greening: strategies for a sustainable world' *Harvard Business Review*, Vol. 75 No. 1 pp. 67–76.

[13] Higher Education Partnership for Sustainability, (2003), *'Accounting for Sustainability: Guidance for Higher Education Institutions'*, November, pp. 7.

[14] Holmes, S. L. (1976), 'Executive perceptions of corporate social responsibility', *Business Horizons,* Vol. 19 pp34–40.

[15] Holt, D. (2003), 'The role and impact of the business school curriculum in shaping environmental education at Middlesex University', *International Journal of Sustainability in Higher Education*, Vol. 4 No. 4, pp. 324–343.

[16] Idowu, S. O. and Towler, B. A. (2004), 'A comparative study of the contents of corporate social responsibility reports of UK companies' *Management of Environmental Quality: An International Journal,* Vol. 15 No. 4 pp. 420–437.

[17] Idowu, S. O. and Papasolomou, I. (2007), 'Are the corporate social responsibility matters based on good intentions or false pretences? An empirical study of the motivations behind the issuing of corporate social responsibility reports by UK companies', *Corporate Governance: The International Journal of Business in Society,* Vol. 7 No 2 pp. 136–147.

[18] International Finance Corporation www.ifc.org

[19] Jensen, M. C. and Meckling, W. H. (1976), 'The theory of the firm: managerial behaviour, agency cost and ownership structure', *Journal of Financial Economics*, Vol. 3 No. 4 pp. 305–360.

[20] Lindblom, C. K. (1994), 'The implications of organisational legitimacy for corporate social performance and disclosure' *Critical perspectives on Accounting Conference*, New York.

[21] Mahapatra, S. (1984), 'Investor reaction to corporate social accounting' *Journal of business, finance and accounting*, Vol. 11 No. 1 pp. 29–40.

[22] Maltby, J. (2004), 'Hadfield Ltd: its annual general meetings 1903–1939 and their relevance for contemporary corporate social reporting' *The British Accounting Review,* Vol. 36 No. 4 pp. 415–439.

[23] Noeke, J. (2000), 'Environmental management systems for universities - A case study' *International Journal of Sustainability in Higher Education*, Vol. 1 No. 3, pp. 237–251.

[24] Norris, G. and O'Dwyer, B. (2004), 'Motivating socially responsive decision making: the operation of management controls in a socially

responsive organisation' *The British Accounting Review,* Vol. 36 No. 2 pp. 173–196.

[25] Ricart, J. E., Rodriguez, M. A. and Sanchez, P. (2005), 'Sustainability in the boardroom: An Empirical examination of Dow Jones Sustainability World Index leaders' *Corporate Governance: The International Journal of Business in Society*, Vol. 5, No.3 pp. 24–41.

[26] Senge, P. M. and Carstedt, G. (2001), 'Innovating our way to the next industrial revolution', *Sloan Management Review*, Vol. 42 No. 2 pp. 24–38.

[27] Spicer, B. H. (1978), 'Investors, corporate social performance and information disclosure: an empirical study', *The accounting review*, Vol. 53 No. 1 pp. 94–111.

[28] Suchman, M. C. (1995), 'Managing legitimacy: strategic and institutional approaches' *Academy of Management Review* Vol. 20 pp. 571–610.

[29] van Weenen, H. (2000), 'Towards a vision of a sustainable university', *International Journal of Sustainability in Higher Education*, Vol. 1 No. 1 pp. 20–34.

[30] Vogl, A. J. (2003) 'Does it pay to be good? *Across the Board*, New York, Jan/Feb. pp. 16–23.

[31] Wals, A. E. J. and Jickling, B. (2002), '"Sustainability" in higher education: From doublethink and newspeak to critical thinking and meaningful learning', *International Journal of Sustainability in Higher Education*, Vol. 3 No. 3, pp. 221–232.

[32] Wright, T. S. A. (2002), 'Definitions and frameworks for environmental sustainability in higher education' *International Journal of Sustainability in Higher Education,* Vol. 3 No. 3, pp. 203–220.

Review of on-site and communal water and sanitation systems for remote communities

A. Perks & T. Johnson
R.V. Anderson Associates Limited, Ottawa, Canada

Abstract

Many small communities across Canada rely on on-site or communal water and wastewater systems to meet their needs, and several factors are likely to reinforce this direction: a) the reduction in grants available from senior levels of government to assist small communities with capital upgrades; b) the emergence of new small scale technologies for water and wastewater treatment that can be cost-effectively applied at the small community level, reducing the need for costly underground piping networks; and c) regulatory pressures to adopt full cost pricing that will force small communities to seek lower cost solutions.

Servicing costs for small communities may be significantly reduced and still provide acceptable and comparable levels of service, as well as employment opportunities within the communities. Because many significant problems and / or failures in on-site and communal water systems have been attributed to inadequate O&M, it is unlikely that more complex technologies requiring higher levels of expertise will represent a sustainable solution. On-site and communal systems may be a more sustainable solution for smaller communities, perhaps using contract O&M services, and should be carefully considered.

Keywords: water, wastewater, onsite, policy, community servicing.

1 Introduction

Throughout Canada, small communities are typically served by on-site or communal water and wastewater systems where it is impractical to construct a centralized system due to the high cost and/or low density of population. According to Environment Canada's MUD survey in 2001, more than three million rural homes and buildings are not connected to municipal systems. Dalhousie University has reported that more than 50% of the population of

WIT Transactions on Ecology and the Environment, Vol 108, © 2008 WIT Press
www.witpress.com, ISSN 1743-3541 (on-line)
doi:10.2495/EEIA080271

Nova Scotia depend upon on-site sewage systems, principally septic tanks and leaching beds.

Within the National Capital Region itself, there are 26 villages within the newly amalgamated City of Ottawa that still rely upon on-site or communal facilities to meet their needs; again, mostly private wells and septic systems. At a recent workshop for rural residents held by the City of Ottawa in March 2007, these rural and village residents expressed satisfaction with their services, and rural homeowner groups such as the Carleton Landowner's Association clearly advocated for decentralized water and sewage systems for village scale communities.

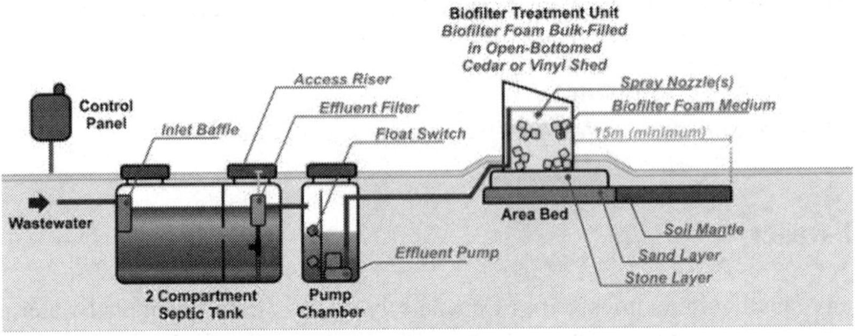

Figure 1: Modern septic system.

Conservation Authorities established on a watershed basis are also playing an important role. The South Nation Conservation, for example, monitors water quality and the operation and maintenance of on-site and communal systems within the watershed – 72 small communities ranging in size from a few houses to small villages of a hundred or more houses. South Nation Conservation provides all inspections of homes or businesses that construct private services. This centralized service ensures consistent standards, and a watershed approach to managing wastewater. Fees are charged for inspections under this program.

2 Managing small community infrastructure

2.1 Servicing strategies

Small and remote communities are often characterized by low growth rates and a need for a complete community water and wastewater servicing solution (compared to a development phased approach). The different infrastructure servicing arrangements can be categorized as follows:

- Category I – On-site wells and septic tanks owned and operated by the residents;
- Category II – Cluster or decentralized public systems (public or private); and
- Category III – Centralized public systems involving complete piped networks.

The simplest technology that meets the appropriate operational standards will usually result in the lowest life cycle costs. Illustrated below is a home scale water treatment unit.

New technologies are now available to enhance the performance of on-site systems, such as filters and aerobic units installed between the septic tank and the leaching bed to prevent bed failure, to remove BOD, suspended solids, nitrates and phosphates, to disinfect effluents, and to accommodate more difficult soils and topography. The next level of treatment devices, including Biological Contactors and Aerobic Treatment package plants are available to handle small communities and clusters of homes. The Canadian Water and Wastewater Association recently prepared a summary of current technologies related to small scale water and wastewater treatment, including equipment for adsorptive media, filtration, disinfection, nutrient removal, and reverse osmosis.

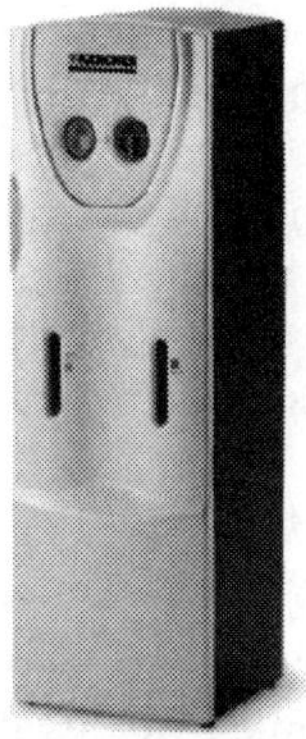

Figure 2: Home water treatment system.

2.2 Asset management

The capital cost of central water supply and wastewater infrastructure in urban areas is typically in the range of $25,000 per house, including treatment plants and piped sewage collection/water distribution infrastructure. These urban systems are provided in urban areas where the lot sizes are relatively small (i.e. 15–20 m widths). Larger rural lot communities are usually serviced by means of private wells and septic systems due to the high cost of the underground water and sewer networks. A simple asset management model for an urban system, broken down into operational cost components, shows how a typical capital investment of $25,000 per connection would break down.

Assuming a water consumption of 375 l/c/d, then these costs would be reflected in a water rate of about $3.88 per m^3 of water consumed.

Not many Canadian municipalities are in fact spending $2,125 per connection per annum at present. Discounting the system renewal and debt repayment costs above brings the annual O&M cost to about $1,250 per annum, not far off a typical homeowners water and sewer bill in most municipalities.

Table 1: Asset management model.

Capital asset value (CAV) per person	$25,000
Annual O&M cost @ 5% of CAV	$1,250
Treatment and purification cost @ 30% O&M	$375
Distribution and collection cost @ 30% of O&M	$375
General and administrative costs @ 40% of O&M	$500
Renewal of short life assets @ 2% of CAV per annum	$500
Debt repayment @ 30% of annual O&M Cost (maximum)	$375
Total annual operating cost	$2,125 or 8.5% of CAV

Much of Canada's water and sewage infrastructure was financed with the assistance of Federal and Provincial grants to assist municipalities with capital upgrades, and as a result many smaller communities have centralized systems offering a high level of service in terms of flows, pressures and capacities. But over the last decade or so, these capital grants have been significantly reduced through a process of downloading and cost cutting. The result has been the emergence of an "infrastructure deficit" - slowly deteriorating assets that are beginning to wear out and break down. Smaller, more remote communities are particularly vulnerable, and are seeking lower cost systems, easier to operate and maintain in the long term.

2.3 System performance

Anticipated cost increases could be at least partially offset by reducing financial losses associated with unaccounted for water and inflow/infiltration. For example, water production and billing records for small communities often indicate a high unaccounted for water rates (30–40%), well beyond acceptable standards for small municipalities (15%). On the other hand, the extra inflow and infiltration entering through leaks and cross-connections and must be treated at the wastewater treatment plant, needlessly reducing the communities ability to handle new development. Both of these represent significant lost revenue and additional operational costs to small communities – as much as 25% of their annual operating budget. Therefore aggressively working to eliminate these losses would help offset rate increases over the next few years.

The development of simple performance indicators or benchmarks could be used to assess the performance of on-site and communal services, such as:

- Percentage of well water tests that comply with standards;
- Percentage of septic tank systems that are inspected each year;
- Percentage of lagoon capacity remaining;
- Percentage of failed leachate beds in the community; and
- Percentage of unplanned service interruptions.

Operators of small systems can establish target values for the performance indicators and determine the system's performance over time, thereby anticipating problems and identifying corrective actions.

3 Operations and maintenance costs

3.1 Water supply

The Ontario Water Works Association used to survey water rates and operations and publish an annual report, but the last one was in 1999, and indicated an annual O&M cost of $412 per connection for water supply for communities with less than 1,000 accounts. This represents about $500 per account in 2007 dollars, or approximately $.91 per m^3 ($910 per ML) of water used, assuming 550 m^3 per year per account.

The Canadian Water and Wastewater Association recently contacted a number of municipalities and agencies to gather data and information on small water and sewage system costs.

The CWWA found that many small communities across Canada still rely on on-site water and sewage systems. The costs of operating these systems varies widely, but on average the operators reported spending about $444 per connection per year on water systems, and $313 per connection per year on wastewater. Assuming a water use of 240 m^3 for a typical household (ie connection), this translates into $1.85 per m^3 ($1850/ML) for water and $1.30 per m^3 ($1304/ML) for wastewater.

In Nova Scotia, the Halifax Regional Water Commission operates a number of small water systems. Operational costs for some of their small systems range from about $800 per ML for a system with 350 customers, to $14.40 per m^3 ($14,400 per ML) for a system serving only 11 customers. The high costs also reflect specific water quality issues to be addressed.

The Ontario Ministry of Municipal Affairs and Housing operates the Municipal Performance Measurement Program (MPMP), which surveys large and small municipalities regarding their operating costs. The most recent compilation for 2003 indicates that for municipal water supply systems serving less than 5,000 persons, operations costs of water produced were $1.03 per m^3 ($1028 per ML – see Figure 3).

The economies of scale are also evident from this figure, as larger municipalities (greater than 40,000 in population) report operating costs as low as $.39 per m^3 ($386 per ML), whereas for smaller communities the unit cost of producing drinking water is much higher.

It is clear that many factors affect the cost of providing drinking water. Chief among these would be the raw water quality and what has to be done to bring the water to an acceptable level of finished water quality. Another important element is the topography; systems able to make use of gravity to supply and distribute water have a natural economic advantage over those that have to pump and re-pump water significant distances and elevations.

The age and condition of the facilities is also an important determinant of production cost – for example unplanned maintenance on an emergency breakdown basis may increase the operating costs significantly - by as much as 15 times if a major component breaks down and damages other parts of the system.

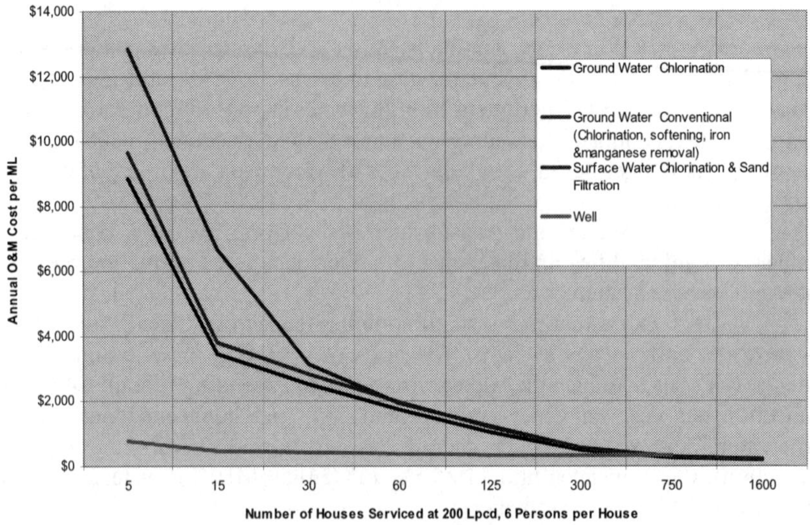

Figure 3: Annual O&M cost for megalitre – water treatment.

Given that up to 85% of the cost of a water system can be represented by the distribution piping, ongoing deterioration and replacement costs can drive the cost of supplying water up to unaffordable levels.

Cost factors may be even more unpredictable for remote communities of several hundred people or less. The cost models currently in use by Indian and Northern Affairs Canada for remote communities indicate that supplying small remote communities can be as much as 40–50 times the cost of operating piped systems for larger urban communities in the 5,000–10,000 population range.

Typically though, an on-site drinking water system comprising a well, appropriate filters (for solids removal) and UV or ozonation disinfection will cost to the order of $300/year for filter replacement and energy consumption. Added point-of-entry systems can be applied for softening, iron or manganese removal and even uranium removal where necessary. These costs can be reduced if dual water systems are built into the houses for non-potable water uses (such as toilet flushing) using untreated water. The Canadian War Museum in Ottawa uses untreated river water for toilet flushing purposes, and city water for all other purposes.

3.2 Wastewater treatment

A similar pattern emerges for wastewater collection and treatment. The operational cost reported in the Ontario MPMP database for communities with a population of less than 1,000 persons is $341 per ML treated, or $0.34 per m^3.

The economies of scale are not as evident, as larger municipalities (greater than 40,000 in population) report operating costs of $.310 per m3 ($310 per ML). This may be due to the fact that many communities rely on simple lagoon systems for treatment that involve relatively little operational costs, even for larger entities.

As mentioned previously, excessive inflow/infiltration entering the systems through leaks and cross-connections can inflate the operations costs rapidly.

The topography is also an important element, as it is with water; systems able to make use of gravity to collect the wastewater have a natural economic advantage over those that have to pump and re-pump the sewage over significant distances and elevations.

The age and condition of the facilities is likewise an important determinant of production cost – for example, cleaning and flushing of sewers that collect grease and debris at many locations and low points are much more costly to operate and maintain.

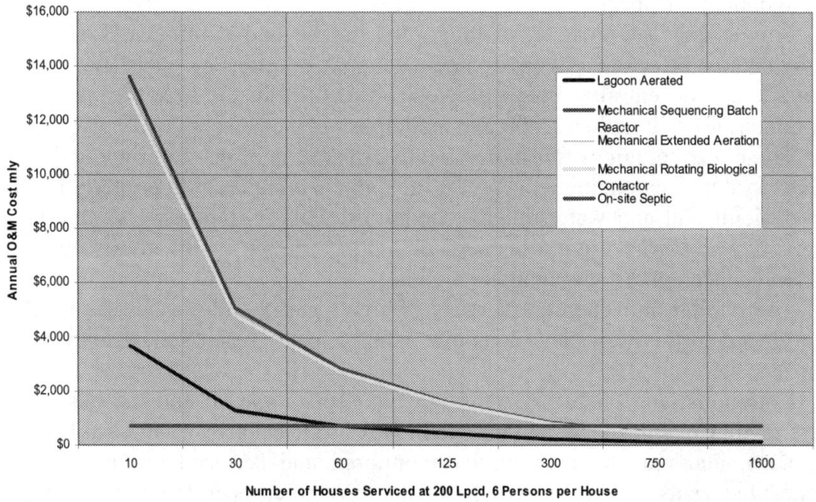

Figure 4: Annual O&M cost comparison – wastewater.

However, perhaps the single most important factor is surface versus subsurface disposal of the effluent. If subsurface disposal can be used, such as for septic tanks, biofilters, etc, the cost is usually an order of magnitude less than the cost of surface water disposal. Therefore, wherever possible, on-site and communal systems relying on subsurface disposal should be considered for small, remote communities.

Figure 5: Typical communal wastewater systems.

4 Summary and conclusions

Much of rural and small town/village Canada (and the USA as well) relies upon private on-site services to meet their water and wastewater needs. Many of these residents prefer their own private, low cost systems and actively advocate against so-called "big pipe" solutions from a cost, environmental and community development perspective.

Simple and low cost technologies for on-site and communal systems can provide cost effective and environmental sound solutions on a sustainable basis. On-site and decentralized technologies appear to offer the least cost service for communities of less than 1,000 in population.

These technologies offer more effective operation of on-site water and wastewater systems, extending their useful life and enabling them to properly function in difficult soil and water quality conditions. This is important, as the cost of moving away from ground disposal of septic effluent to surface discharge, for example, is an order of magnitude higher.

The capital and operational costs of small water and wastewater networks developed in this way may be as low as 33% of the cost of more conventional piped systems.

Communal and on-site systems, such as septic systems and lagoons, require less regular maintenance than more complex electrical and mechanical treatment systems, and may be more easily monitored and serviced through regularly scheduled visits. However, past problems have often been traced to inadequate cleaning and inspection. Operational and management contracts are available to assume responsibility for the maintenance of small systems – which could be incorporated into private cluster services as well. Otherwise, regional or watershed bodies can be contracted to perform the inspection and maintenance functions for a number of communities.

A hierarchy of infrastructure strategies should be considered, beginning with on-site systems, communal and decentralized systems, and finally full piped centralized systems only where dictated by population size and density of development. Where a community is more compact in area with potential well contamination from septic systems, a communal water supply may offer a cost effective solution.

Further benefits may be gained by designing on-site and communal systems for wastewater recycling and reuse in which perhaps 50% of the wastewater can be recycled for use in irrigation and/or toilet flushing, for example. This further enhances the potential cost efficiency and environmental benefits for small community services.

Policies to mitigate the damage from coastal natural disasters: preparing southeastern U.S. coastal communities

J. Pompe
Economics Department, Francis Marion University, USA

Abstract

The combination of densely populated areas, high-valued real estate, and physical vulnerability to severe storms has led to some of the costliest natural disasters in history. In light of concerns that global warming will increase the vulnerability of coastal areas, policy approaches that can encourage sustainable development and aid coastal zone management are important. Policies that protect natural defenses, direct development away from hazardous areas, enforce storm-resistant building codes, and remove perverse incentives, will promote sustainable communities in vulnerable coastal areas.
Keywords: sustainable coastal communities, natural disasters.

1 Introduction

Residents of coastal areas, especially along the southeastern United States coast, have suffered through increasingly costly storms in recent years. The five most intense consecutive storm seasons on record occurred between 1995 and 2000. An unprecedented 4 hurricanes, viz. Charley, Frances, Ivan and Jeanne, damaged Florida communities during 2004. The 2005 hurricane season was the busiest and most costly in United States history, with 28 named storms, 15 of which were hurricanes, including Katrina.

In part, recent storms have become more costly because of rapid population growth in coastal areas, which leads to the construction of more homes and businesses. NOAA's report [1] on coastal population trends in the United States shows that from 1980 to 2003 coastal population increased from 120 to 153 million people, an increase of 28 percent. Projections suggest that another 11 million people will move to coastal counties by the year 2008 for another 7

WIT Transactions on Ecology and the Environment, Vol 108, © 2008 WIT Press
www.witpress.com, ISSN 1743-3541 (on-line)
doi:10.2495/EEIA080281

percent increase. Much of the coastal population growth has occurred on the shorelines of southeastern and Gulf Coast states. Coastal population density of the Southeast region (Florida, Georgia, North Carolina, and South Carolina) increased from 142 to 224 persons per square mile between 1980 and 2003, and is expected to increase to 241 by 2008 [1]. The current southeast region population density is 2½ times the population density of the nation which is 98 persons per square mile.

The combination of densely populated areas, high-valued real estate, and physical vulnerability to severe storms has led to some of the costliest natural disasters in history. The 2004 and 2005 hurricane seasons produced seven of the ten costliest insured losses ever to affect the United States [2]. Katrina was the most costly, causing $40.6 billion in damages and in total, the seven hurricanes caused $79.3 billion in insured losses. Insurance companies, responding to the higher costs, have increased premiums and even withdrawn from some markets. Consequently, there has been growing pressure for increased government intervention to help coastal residents with property insurance costs.

Rising sea level and coastal erosion increase the risk from storms that coastal residents must face. Some are projecting that sea level may increase by a meter or more by 2100 [3], which will inundate some coastal areas and increase the exposure of other areas. In light of concerns that global warming and sea level rise will increase the vulnerability of coastal areas, I consider policy approaches that can encourage sustainable development and aid coastal zone management.

2 Policies to encourage sustainable coastal communities

Correcting human activities that contribute to the creation of more vulnerable coastal communities can help protect coastal residents and their property. For example, human created problems, such as development in hazardous areas, poorly constructed levies, and the human destruction of wetlands contributed to the Katrina disaster. Government can implement various policies to improve the likelihood of sustainable communities.

2.1 Land-use policies to protect natural defenses

Implementing land-use policies that balance development and ecosystem protection will lessen coastal community vulnerability to storm damage. Ecosystem management, such as protection of wetlands, which absorb some of a hurricane's impact, can reduce storm damage. Costanza, et al. [4] estimated that coastal Louisiana wetlands provided $400/acres/year (in 2006 dollars) in storm reduction benefits. Although the Mississippi deltaic plain bordering Louisiana provides valuable storm damage reduction benefits, numerous human activities have destroyed large areas of the coastal marshes. Since 1900, about 4900 km^2 of wetlands in coastal Louisiana have been lost at rates as high as 100 km^2/year.

In order to maintain and enhance Gulf Coast coastal wetlands, the federal government has undertaken a large-scale effort, which includes using dredged sediments to restore wetlands and barrier islands [5]. In addition, the team of researchers that produced *Coast 2050* [6], a year and a half study completed in 1998, recommended that the Mississippi River be re-engineered to imitate natural processes, in order to restore wetlands. The study group estimated that re-diverting the River's flow would cost $14 billion in 1998.

Market-incentive policies such as Wetland Mitigation Banking and the Wetland Reserve Program (WRP) help protect and restore wetlands that can help protect coastal areas. For example, the WRP, which the United States Department of Agriculture's Natural Resources Conservation administers, offers financial incentives and technical assistance to landowners to protect, restore, and enhance wetlands on their property. The WRP, which was established by the 1990 Farm Bill, offers tax incentives and payments to landowners who agree to restore previously converted wetlands. Most WRP sites are in flood prone areas [7].

Efforts to protect barrier islands, which parallel mainland areas, are important for coastal protection. Barrier islands, which absorb wave action and storm surge, provide a buffer against storms and offer a valuable habitat for fish and wildlife. The Federal government attempts to protect barrier islands and discourage development in hazardous areas through the 1982 Coastal Barrier Resource Act (CBRA). Communities in CBRA areas (certain high-risk areas on barrier islands) are restricted from receiving any federal financial assistance such as post-storm reconstruction and erosion control, in addition to National Flood Insurance Program (NFIP) insurance. Unfortunately, a 1992 review found that new development was continuing in CBRA areas and that 9 percent of the residents received NFIP insurance [8]. Increased efforts to discourage development in hazardous areas, especially the barrier islands in CRBA, should be undertaken. Government agencies involved with flood insurance should make it clear that buildings in such areas will not qualify for subsidized insurance. If such a policy is made clear prior to any construction in such areas, less development will take place since property owners risk costs would be higher.

Shoreline erosion, which is caused by natural processes and humankind projects, exposes coastal communities to increased threats. The effects of erosion may be mitigated with land-use policies that protect dunes and natural vegetation, which stabilize shorelines. Some states, for example, prohibit the destruction of vegetation on shoreline areas.

2.2 Directing development

Government can reduce damage from natural disasters by directing development away from undeveloped floodplains and erosion zones, as recommended by the U.S. Commission on Ocean Policy [9 , p. 168]. Although the goal of the 1968-established National Flood Insurance Program (NFIP), which provides subsidized flood insurance, was to direct development away from hazardous areas, it has not been effective. With more than a third of the 6.6 million buildings located in the 100-year floodplains of participating communities built after the NFIP floodplain management plan, the program has not successfully directed development away from the path of floods [10].

Some states implement setback regulations that require new construction to be set back from the shoreline a minimum distance, which may be determined by erosion rate or land elevation. In South Carolina, for example, 20 feet landward of a baseline, drawn at the point of the highest ridge line of the primary dune, is the closest to the shoreline that a structure can be built [11]. Such setbacks will minimize the negative effects of erosive shorelines, sea level rise, and storm damage costs.

In those hazardous areas already developed, government can encourage relocation. For example, the property buyout program, authorized by the 1988 Stafford Act, provides federal dollars to purchase property in floodplains to reduce future expenditures from flood damage [12]. The purchased areas must not be allowed to be developed again, because this would subsidize development.

Titus [13] suggests rolling easements as a way to adapt to rising sea level. Such an approach would have the government purchase a property when sea level rose to a certain level. Because a house is removed as the shore moves toward it, the beach profile would be maintained as the shore migrated inland. A policy of "retreat" from the beach, which may involve purchasing threatened property, will be costly in most coastal areas that are developed. Parsons and Powell [14], for example, estimate that over the next 50 years, the cost of retreat for Delaware beaches would be $291 million in present value terms. An alternative policy of beach nourishment, which would protect real estate from the erosive shoreline, would cost $60 million for the same period.

Policies to limit population growth in hazardous coastal areas may be desirable. Impact fees, for example, could be imposed on new housing to discourage development in hazardous areas. Generally, impact fees, are meant to compensate a local government for the fiscal impact of new housing and therefore may not have much of an impact on limiting development. Limiting building permits can decrease the amount of building activity in an area. By restricting the supply of new housing, population growth may be slowed.

2.3 Storm-resistant construction and engineering projects

Encouraging storm-resistant construction will mitigate hurricane damage. Fronstin and Holtman [15] found that Hurricane Andrew, which struck Florida in 1992, caused less damage in subdivisions with higher average home prices. Although houses were more expensive in higher income areas, because the houses were more storm-resistant, nearby houses suffered less damage from wind-blown debris. A quarter of the $16 billion in insured losses from Hurricane Andrew were attributed to Dade County's failure to enforce building codes [16]. Poor enforcement of building codes has been a problem in many of the southeastern states. At the time of the Gulf coast hurricanes in 2005, building codes were not enforced adequately in Alabama, Louisiana, Mississippi, and Texas [17].

Although building codes have not always been enforced, many southeastern state and county governments are improving building standards in coastal areas. Louisiana passed legislation in December of 2006 requiring the 11 parishes hardest hit by Katrina to begin within 90 days to meet the wind and flood provisions of the International Building and Residential Codes. This building code requires homes built along the Gulf Coast to withstand winds of 130 to 150 miles per hour. In order to meet this standard, builders will rely more heavily on steel and concrete in construction. Construction practices such as hurricane–resistant windows, metal strapping from the foundation to the roof, and houses wrapped with plywood will become commonplace [18]. In addition, houses in floodplains can be built on stilts a story above the ground to reduce storm surge damage.

Engineering technology such as the barriers the British and Dutch built to protect against North Sea storm surge are expensive but may provide an engineering option for some threatened areas. Great Britain built steel barriers across the Thames River in 1983 at a cost of 500 million pounds to protect London from storm surge. The Dutch designed the Oosterschelderkering, a huge barrier completed over two decades ago at a cost of 2.5 billion euros, south of Rotterdam [19].

Shoreline erosion, which averages two to three feet per year along Atlantic coast and six feet per year on the Gulf coasts [20], can exacerbate the threat to coastal communities. Beach nourishment, a process that pumps sand onto the beach, is a generally accepted method to mitigate shoreline erosion, although it is expensive and temporary. The average cost for a cubic yard of sand is approximately $5. Developed states should expect to spend $6 million per mile of shoreline each decade. In 1992, a northeaster washed away most of the sand from a 1980s $51.2 million Ocean City, Maryland project that nourished a 9 mile stretch of beach [21]. However, in coastal areas with costly development, such costs may be justified by the storm damage reduction benefits created by wider beaches.

2.4 Combating global warming

Policies to combat global warming are an integral part of a program to mitigate coastal disasters. The increase in temperatures that will melt polar ice caps and cause an increase in sea level rise will inundate many coastal areas such as Miami Beach, New Orleans, and Hampton, Virginia [22]. In addition, if expectations are correct that global warming will increase hurricane intensity, occurrence, and landfall frequency [23], curbing the actions that contribute of global warming will be doubly important for many coastal residents.

Market-incentive policies offer the best hopes of success for a problem such as global warming. Marketable permits, which set a limit on allowable carbon emissions, restrict greenhouse emissions in a cost-effective manner by allowing trades between polluters. A carbon tax would levy a fee on carbon gas emissions which encourages polluters to decrease their emissions. Both incentive programs encourage least-cost solutions that are preferable to a command-and-control policy. Policies that encourage the protection and planting of forests, which absorb carbon dioxide, will help reduce carbon dioxide emissions. Credit payments for forest protection would therefore help offset the damage from greenhouse gasses.

Policies that encourage energy efficiency, such as energy efficient buildings, will decrease fossil fuel use. Insurance companies could encourage building or rebuilding to meet "green construction" standards [24]. For example, Cementitious Structurally Insulated Panels, a technology championed by the Federation of American Scientists, has wind-resistant cladding and styrofoam cores, that provide high energy efficiency and reduces the amount of wood required for construction.

2.5 Information on vulnerable areas

Information about areas that are physically vulnerable should be disseminated by governments and utilized for development planning, as well as for informing residents. Agencies such as the United States Geological Service, Federal Emergency Management Agency, and the United States Army Corps of Engineers, have identified hazard-prone shorelines, based on pre-storm geomorphic factors [25]. Thieler and Hammer-Klose [26], for example, have created a physical vulnerability index for U.S. coastal areas based on six variables: 1) geomorphology, 2) shoreline erosion and accretion rates, 3) coastal slope, 4) rate of relative sea-level rise, 5) mean tidal range, and 6) mean wave height. Areas with high-energy coastlines, a low regional coastal slope, and where the major landform type is a barrier island, are the most vulnerable, generally [26]. Based on this analysis, sections of the mid-Atlantic coast (Maryland south to North Carolina) and northern Florida, are some of the areas of greatest vulnerability.

In addition, some areas are more susceptible to storm strikes. The greatest likelihood of severe damage from hurricanes is along the coastlines of the southeastern Atlantic and Gulf of Mexico states, where 112 major hurricanes have struck between 1851 and 2006 [27]. Although predictions of where the next major storm will hit are problematic, clearly some locations are more prone to suffer from storms than are others. Historically, 39 percent of all major hurricanes in the United States have battered Florida, and 71 percent of category 4 or 5 hurricane strikes have pummeled either Florida or Texas.

2.6 Removing perverse incentives

Unfortunately, numerous well-meaning government policies, such as the NFIP, that subsidize coastal development contribute to the problem. Removing perverse incentives that encourage development in hazardous coastal areas will reduce storm damage costs and create more sustainable communities. The NFIP, which provides subsidized flood insurance, was meant to ensure that most residents in high hazard areas were insured and to direct development away from the most flood-prone areas. It is clear that the program has accomplished neither goal. Indeed, subsidizing insurance, which encourages development in hazardous areas, has been counter-productive. In addition to subsidizing coastal development in hazardous areas, the NFIP is not actuarially sound, insurance penetration rates are low, and repetitive-loss properties absorb a large percentage of payments [28, 5].

Government should minimize involvement in property insurance markets in order to avoid crowding out more efficient market solutions. Instead, the government should facilitate the development of private market solutions that will insure that those who choose to live in high-risk areas, will pay the cost of such a choice. A benefit of privately provided insurance is that companies base rates on risks involved, and thus rates will be high if risk is high. Risk-based rates encourage investment in cost-effective risk mitigation measures, such as fortified buildings, because companies will lower rates on construction that is less likely to be damaged. In addition, because rates will be highest in the most hazardous areas, residents will be less likely to locate in such areas. If insurance rates become too much of a burden for low-income families living in hazardous areas, rather than offering subsidized flood insurance, a voucher similar to food stamps could be issued. Although the homeowner would pay a risk-based premium, policy-makers could base the magnitude of the voucher on the income and assets of the resident.

In addition, government lowers the cost of residents who choose to locate in hazardous areas by paying some or all of the cost of beach nourishment, public infrastructure replacement, and disaster relief. For example, the federal government pays for much of the cost of many beach nourishment projects. Although benefits may be greater than the costs of such projects in areas where valuable property is threatened, government provided funding encourages development. Disaster relief and other government programs (such as subsidized insurance rates) that bail out people from all disasters, however well intentioned, encourage more risky choices that lead to repeat disasters. By subsidizing coastal areas with such programs, government creates a “moral hazard” which makes a full-blown disaster more likely. A moral hazard is created when an individual has no incentive to guard against a risk if already protected from the risk. Indeed, individuals have little interest in avoiding building in a hazardous area if someone else incurs the cost.

If government chooses to help residents withstand the damage from severe storms, policies that encourage spending on mitigation measures would be the more efficient than lowering insurance premiums. For example, in 2007, South Carolina legislators passed new legislation that offered tax credits for building supplies that made homes storm resistant and insurance premium discounts for residents who made homes storm resistant [29]. Such policies encourage storm resistant homes, which will reduce damage costs, and reward choices that decrease damage costs from hurricanes.

3 Summary

Residents in many coastal areas are subject to natural threats such as hurricanes, northeasters, and coastal erosion that may threaten community sustainability. Discontinuing programs such as the NFIP that create perverse incentives which encourage development in hazardous areas, will create more efficient and equitable results. In order to create sustainable development, land-use policies that protect natural areas such as wetlands, will balance development with ecosystem protection. Factoring the storm damage reduction benefits of ecosystems, such as wetlands, into benefit-cost analyses would encourage protection of valuable natural defenses. Other solutions discussed above include providing better information on vulnerable areas, encouraging storm-resistant construction, examining engineering technology, and implementing policies to combat global warming.

Cicin-Sain and Knecht [30] and others have suggested that integrated coastal management (ICM), which emphasizes the interrelationship between inland areas, coast, and ocean, is necessary to protect coastal areas. ICM requires interdisciplinary, intergovernmental, and international cooperation on policy solutions. Population growth is likely to continue along coastlines worldwide, increasing the importance of considering the problems that coastal communities must address.

References

[1] Crossett, K.M., Culliton, T.J., Wiley, P.C., & Goodspeed, T.R., *Population Trends Along the Coastal Untied States: 1980-2008*. National Oceanic and Atmospheric Administration, NOAA's National Ocean Service, Special Projects: Silver Spring, MD, 2004.

[2] Insurance Information Institute. Hurricane and Windstorm Deductibles. 2007. www.iii.org/media/hottopics/insurance/hurricanwindstorm/

[3] Kerr, R.A., Pushing the scary side of global warming. *Science*, **316 (5830)**, pp. 1412–1415, 2007.

[4] Costanza, R., d'Arge, R., de Groot, R., Farber, S., Grasso, M., Hannon, B., Limburg, K., Naeem, S., O'Neill, R. V., Paruelo, J., Raskin, R. G., Sutton, P., & Van der Belt, M., The value of the world's ecosystem services and natural capital. *Ecological Economics*. **25 (1)**, pp. 3–15, 1998.

[5] Day, J.W., Jr., et al., Restoration of the Mississippi Delta: lessons from Hurricanes Katrina and Rita. *Science* 23, **315, (5819)**, pp. 1679–1684, March 2007.

[6] Louisiana Coastal Wetlands Conservation and Restoration Task Force and the Wetlands Conservation and Restoration Authority. *Coast 2050: Toward a Sustainable Coastal Louisiana.* Louisiana Department of Natural Resources. Baton Rouge, La., 1998.

[7] National Resources Conservation Service. *Farm Bill 2002: Wetlands Reserve Program.* United States Department of Agriculture, Washington, D.C., September 2004.

[8] United States General Accounting Office (US GAO). *Coastal barriers: Development occurring despite restrictions.* Washington, D.C., 1992.

[9] U.S. Commission on Ocean Policy. *An Ocean Blueprint for the 21st Century.* Washington, DC, 2004.

[10] Burby, R.J., Flood insurance and floodplain management: the U.S. experience. *Environmental Hazards* **3, 3**: pp. 111–122, 2001.

[11] South Carolina Beachfront Management Act. SC 49-39-250, 1988.

[12] Pasterick, E. T., The National Flood Insurance Program. *Paying the Price: The Status and Role of Insurance Against Natural Disasters in the United States*. Howard Kunreuther and Richard J. Roth, Sr., eds., Joseph Henry Press: Washington, DC, 1998.

[13] Titus, J. G., Rising seas, coastal erosion, and the takings clause: how to save wetlands and beaches without hurting property owners. *Maryland Law Review*, **57, (4),** pp. 1279–1399, 1998.

[14] Parsons, G. R. & Powell, M., Measuring the cost of beach retreat. *Coastal Management*, **29**, **(2),** pp. 91–103, 2001.

[15] Fronstin, P. & Holtman, A.G., The determinants of residential property damage caused by Hurricane Andrew. *Southern Economic Journal* **61**, pp. 387–97, 1994.

[16] Building Performance Assessment Team, *Preliminary Report in Response to Hurricane Andrew, Dade County, Florida*. Federal Emergency Management Agency: Washington, 1992.

[17] Burby, R. J., Hurricane Katrina and the paradoxes of government disaster policy: bringing about wiser governmental decisions for hazardous areas. *Annals of the American Academy of Political and Social Science* **604**, pp.171–192, 2006.

[18] McLeister, D. Lessons from Katrina: better building, codes, materials. July 7: 2, 2007. www.hgtvpro.com/hpro/nws_dstr_huric_torndo/article/0,2624, HPRO_265224503255,00.html

[19] Barnet, J. & Hill, K., Design for rising sea levels. *Harvard Design Magazine* **27**, Fall 2007/Winter 2008.

[20] Baish, S., Dunn, S. & Friedman, R., Coastal erosion: evaluating the risk. *Environment*, **22, (7),** pp. 36–45, 2000.

[21] Trebanis, A. C., Pilkey, O. H. & Valverde, H.R., Comparison of beach nourishment along the U.S. Atlantic, Great Lakes, Gulf of Mexico, and New England shorelines. *Coastal Management,* **27**, pp. 329–340, 1999.

[22] Mazria, E. & Kershner, K., *Nation under Siege: Sea Level Rise at our Doorstep*. The 2030 Research Center, 2007.

[23] Trenberth, K. E. Warmer oceans, stronger hurricanes. *Scientific American*, July, pp. 45–51, 2007.

[24] Mills, E. & E. Lecomte, From Risk to Opportunity: How Insurers Can Proactively and Profitably Manage Climate Change. Ceres: Boston, MA, 2006.

[25] Young, R., The high cost of subsidized coastal development. *Geotimes* February 2006. /E:/flood%20insurance/young.subcoast.2006.htm#author

[26] Thieler, E. R. & Hammar-Klose, E.S., National Assessment of Coastal Vulnerability to Sea-Level Rise: Preliminary Results for the U.S. Atlantic Coast. U.S. Geological Survey, Woods Hole: Massachusetts, 1999.

[27] Blake, E. S., Rappaport, E.N., & Landsea, C.W., *The Deadliest, Costliest, and Most Intense United States Tropical Cyclones from 1851 to 2006 (and Other Frequently Requested Hurricane Facts)*. NOAA Technical Memorandum NWS TPC-5. Miami: National Weather Service, 2007.

[28] Jenkins, W. O., Jr. *Federal Emergency Management Agency: Challenges Facing the National Flood Insurance Program.* GAO-06-174T. Washington, DC: U.S. General Accounting Office, 2005.

[29] Faber, J., See how the State's new coastal insurance law could affect your wallet. *The Island Packet*, Hilton Head Island, SC, June 18, 2007. www.islandpacket.com/news/local/story/6554778p-5833850c.html

[30] Cicin-Knecht, B. & Knecht, R., *Integrated Coastal and Ocean Management: Concepts and Practices*. Island Press: Washington, D.C., 1998.

In the absence of their men: Women and forest management in the Mid-hills of Nepal

K. Giri[1], B. Pokhrel[2] & I. Darnhofer[1]
[1]*University of Natural Resources and Applied Life Sciences, Vienna, Austria*
[2]*Programme Manager, Nepal Swiss Community Forestry Project, SDC, Nepal*

Abstract

In Nepal, the management of community forests is based on the participation and decision making of forest users. The premise of its success is the involvement of the real users in forest conservation and management. The Nepal Forest Laws identify women as key forest users and underline the importance of their participation in community forest management. However, given the socio-cultural setting and the prevailing patriarchy, fostering women's active participation remains a challenge. Women are traditionally limited to the private sphere and men tend to look after the responsibilities in the public sphere. However, the increasing trend of male outmigration observed in the Mid-hills may offer a window of opportunity for women to become more involved in the public sphere and thus, be able to have a decisive influence in forest management. This paper investigates the factors that have increased the participation and decision-making level of women in two community forest user groups. Data were collected through focus group discussions, informal discussions and interviews with key informants. The results suggest that key factors that encourage women to take an active role in the management of community forests are: degraded forests hampering the women to fulfil their duties (supply of firewood, grass, etc.), previous experiences with women's groups to increase their self-confidence, an unsatisfactory flow of information and men's full support. Given the high prevalence of male outmigration in the Mid-hills of Nepal, these results are relevant to formulate policies and strategies that foster women's empowerment.
Keywords: community forestry, community forest user group, male outmigration, left-behind women, participation, decision-making, focus group discussion.

WIT Transactions on Ecology and the Environment, Vol 108, © 2008 WIT Press
www.witpress.com, ISSN 1743-3541 (on-line)
doi:10.2495/EEIA080291

1 Women's participation in community forestry

Promoting the participation in decision-making of the less vocal and less powerful has remained orthodoxy for development work. In the management of natural resources such as forests, the emergence and institutionalization of participatory programmes has taken various forms under such umbrella terms as social forestry, collaborative forest management or community forestry.

The concept of local people's involvement in the management of natural resources is not new. What might be new is that to empower local people, structured models of participation are used, built around specific decentralized policy frameworks. Community forestry is one of the highly acclaimed participatory programmes in Nepal that implements the principles of decentralization [1]. It aims to cover the basic needs for forest products of the local people by ensuring their participation through the formation of a community group, widely known as "community forest user group" (CFUG). CFUGs are cohorts of users of a certain forest at the local level (neighbourhood, ward or village) that enjoy use rights, after the forest has been handed over from the state to the community. Each CFUG is governed by an Executive Committee that acts on behalf of the General Assembly of all members.

Participation is a dynamic process through which stakeholders influence decisions and share control over the resources and the development initiatives that affect them [2]. Participation is defined in its narrowest sense in terms of nominal membership and in the broadest sense as a process in which the voice of the disadvantaged (e.g. women) is heard and they thus, influence decision making [3]. According to Agarwal's [3] "ladder of participation", women's participation in a CFUG is "passive" if they may get some information about community forest management but lack any opportunity to make choices or influence the decisions, whereas an "active" participation is characterised as women voicing their views, whether solicited or not, and their ability to actively and directly influence the different initiatives of the CFUG.

Whereas the participatory approaches and decentralized policies of community forestry promise inclusion by creating spaces to exercise decision-making and equitable development, claims to women's participation and decision-making into such "participatory" processes has remained mostly a rhetoric [4, 5]. Indeed, evidence suggests that women's involvement in community forestry has mostly been "passive", represented in the form of women's household entitlement to CFUG membership [2, 3, 5]. As such, women are often simply position holders without the possibility to influence decision-making.

Empirical evidence indicates various factors that constrain women's participation in community forestry. Some argue that the socio-cultural context of Nepalese society and the existing local power structure that provides more power to men can lead to "participatory exclusion" of women [3, 5, 6]. The influence of the socio-cultural context may be maintained through the resistance from village men, based on expectations regarding the behaviour of women during public forestry meetings [7–9]. Also women's needs and aspirations regarding the timings of forest meetings might not be taken into account, nor their lack of self-confidence in a public setting [6, 8]. As such, traditional gender roles assigning different responsibilities to women and men can restrict women's access to natural resources and frequently excludes them from decision-making in community forest management.

While the effect of the socio-cultural context has been reported to affect women's inclusion and influence on decision-making in community forestry, this social-cultural context is not static but undergoes continuous negotiations, adaptations and changes. Male outmigration has been widely reported as one mediating factor that can bring forth negotiations and social transformations by (re)structuring traditional gender roles, thus leading to increased access to resources, greater decision-making powers [10–12] and women's active involvement in community development activities and farming [12, 13].

Given the "passive" state of women's participation in community forest management and the potential of male outmigration to mediate changes in social relations, this paper aims to explore how rural women's participation and decision-making in community forest management has been affected by male outmigration. It also offers indications of the impacts of women's participation and decision-making in community forest management and the continuing constrains and challenges they face.

2 Methodology

2.1 Site selection

The study was conducted in the Mid-hills, a mountain range that crosses Nepal from east to west, between the Himalayan range in the north and the Ganges River plain in the south. The altitude of the Mid-hills varies between 1,000 and 3,000 m. The Kavre district, some 70 km east of Kathmandu, was selected as livelihoods rely mostly on subsistence agriculture, livestock farming and forest resources [14]. Also, Kavre district boarders Kathmandu and is well-connected to other major towns such as Dhulikhel and Banepa. Therefore, many men come to these cities either for study, work or business. In addition, Central Bureau of Statistics [15] reports many of the men from Kavre district go to other countries such as India, Malaysia and Saudi Arab for employment.

For this study, two CFUGs with a high rate of male outmigration were selected. As official statistical data on migration is inadequate to grasp the accounts of outmigration within Nepal, outmigration levels in Kavre district were assessed through discussions with key informants from District Forest Offices, District Development Committees (a local administrative unit acting at district level), range posts, and NGOs. This provided a preliminary list of areas within Kavre with particularly high rates of male outmigration. Six CFUGs were then visited to check the rate of male outmigration and other characteristics of the CFUG through discussions with members of the Village Development Committee, school teachers, as well as members of the CFUG and its executive committee. Finally, two CFUGs – Chande Majuwa and Katunje Pakha – were selected, as both had a high rate of male outmigration and an active participation of women in the CFUG. Also, the two CFUGs are similar in other important aspects, such as access to markets, income from the CFUGs and exposure to tours and trainings.

2.2 Data collection

Primary data was collected between November 2007 and January 2008 through focus group discussions, individual interviews and participant observation.

Three focus group discussions were carried out with ten women in each CFUG. Each focus group discussion took about two hours. The main issues discussed were the factors that motivated women to participate in the management of the community forest (which were ranked in order of importance in each CFUG), the changes that took place after women started to participate, and women's perception regarding men's attitude towards women's participation in these CFUGs.

Furthermore, informal discussions with male members of the CFUG were conducted to assess their perception of women's involvement in community forest management. Additionally, individual interviews with key informants such as the school teacher, forest rangers and local tea-shop owners were conducted to explore the issues of forest condition and management. The data was transcribed, analysed qualitatively and triangulated with secondary information obtained from the minutes, constitutions and operational plans of the CFUGs.

3 Results and discussion

3.1 Factors influencing women's participation in the management of the community forest

Forest management in both CFUGs started about 25 years ago, through the reforestation project of the Nepal Australia Forestry Project. Both community forests were formally handed over to the CFUG about 15 years ago. At that time, women's participation was predominantly passive. Male CFUG members held meetings and took decisions while women were barely – if at all – informed about the timing and/or outcome of these meetings. Women were unaware of the functioning of the CFUG and the potential benefits they could gain from the use of CFUG funds. However, in the last five years, women's awareness and stake in forest management has increased, so that now their participation in decision-making can be described as active.

As the main factors that allowed for this increased participation and active engagement in the decision-making within the CFUG, the women in the focus groups stated that collecting forest products is their responsibility, and that through their increased awareness of the importance of the CFUG and their confidence in their own abilities to manage the CFUG, they started to take a more active role in the management of their community forest (see Fig. 1).

3.1.1 Forest products and water are women's responsibility

Since in Nepal the collection of forest products such as fuelwood, fodder, grass and bedding material is mainly women's responsibility [4, 9], women in both

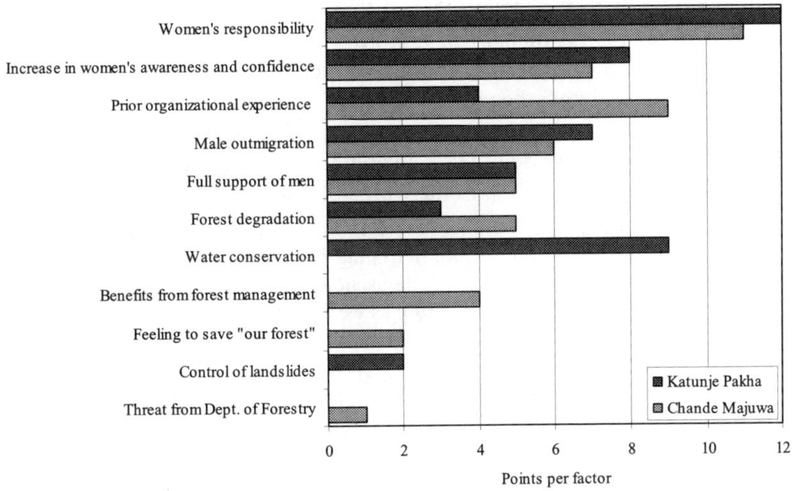

Figure 1: Weighed ranking of factors that motivated women to increase their participation in the management of the community forest. [Note: Each of the 10 women participating in the focus groups was given 5 points to distribute among the factors listed. Not all factors were listed in both CFUGs.]

CFUGs started to face problems in meeting their household requirements as the state of the community forest degraded. Pressured to meet their household duties, women started to sneak into nearby community forests or national forest to collect forest products. However, these were farther away, so that the women had to spend more time to collect the forest products. Also, if the women were caught stealing the forest products from other CFUGs or a national forest, they had to face penalties for misbehaviour and public shame. Securing a regular flow of forest products, therefore, became a core issue for the women, encouraging a more active participation in their own CFUG.

3.1.2 Women's increased awareness and confidence

The adult literacy programmes conducted by the Village Development Committee in both CFUGs provided a venue where women could sit together and learn in groups. This opportunity for information exchange made them more aware of the benefits they could potentially derive from forest management, such as planting medicinal plants in the forest to generate an income, or using CFUG funds from wood sales to address community problems.

3.1.3 Prior experience in organization

Approximately four years ago, women had the opportunity to get involved in some other organizations. In Chande Majuwa, women started a 'saving and credit scheme' where each woman had to contribute 100 Nepalese Rupees (Rs.) per month. This allowed the women to set up a revolving fund that was used to solve the problems of member households in times of need. This experience provided women with the feeling that, if they organized themselves, they could solve their problems on their own, i.e. they did not always have to depend on their husbands or on another male household member.

This experience strengthened the women's feeling of self-confidence and showed them the potential benefits they could derive from a successful organization. It also increased men's awareness and acceptance that women can successfully lead organizations. In the words of a woman in the focus group:

> "Before, women in these villages were limited to performing assigned duties within their household only. But after being involved with the saving group, I also took on responsibilities of my household just like my husband. This has increased my self-esteem in my family as well as in society."
>
> Focus group discussion, Chande Majuwa CFUG

Women in Katunje Pakha participated in a programme for children and women, called DOCAW, initiated by the Katunje Village Development Committee. Among others, the DOCAW provided training to raise the women's awareness of their legal rights. Participation in this training has enhanced women's knowledge and their self-confidence:

> "Before, I did not know anything. Participation in DOCAW made me aware about my own rights as a woman. It has also increased my self-confidence and capability to voice my concerns in public meetings."
>
> Focus group discussion, Katunje Pakha CFUG

3.1.4 The high rate of male outmigration

The former Executive Committee of the Chande Majuwa CFUG was a men-only committee. When they made decisions about forest regulations, women tended not to receive any information about the timing of meetings or the decisions taken:

> "Earlier we did not even hear about meetings. Men used to do that. They also did not use to share information. We didn't even know when the forest was opened and closed. We thought that it was only men who should held meetings and make decisions."
>
> Focus group discussion, Chande Manjuwa CFUG

In Katunje Pakha, women were formally included in the initial Executive Committee, but men monopolized the decision-making, so that the women ended up not participating in the meetings.

When the rate of male outmigration increased, this led to a lack of guidance within the CFUG. Indeed, in Chande Majuwa most of the male members of the Executive Committee left for cities in search of better employment. Thus, the men were no longer present and able to provide the time required to solve the various problems in the community forest. As a result illegal tree felling and forest encroachment was rampant in both CFUGs. In Katunje Pakha, forest degradation led to issues of water scarcity and landslides, which were a core concern of the women.

3.1.5 Full support of village men

Given their inability to cope with the rampant forest degradation, combined with an increased confidence in women's ability, men in both CFUGs finally encouraged women to come to the fore and take part in decision-making on protection, management and use of the community forest. In both CFUGs, women perceived that male members fully supported their engagement. Men thought that if women participated in decision making, introducing their perspective and concerns, the forest would be better cared for. Indeed, since it is mostly the women who go to forests to collect forest products, they tend to be the most knowledgeable [7, 9] about the forest condition, areas of illegal felling and even the illegal encroachers. In Chande Majuwa – combined with the outmigration of the male members of the Executive Committee – this led to the formation of an all-women Executive Committee. In Katunje Pakha the women's share was increased to 50% of the Committee members (up from 10% about four years ago).

3.2 Family composition and remittances as mediating factors

A left-behind woman has to cope with new responsibilities in the absence of her husband [11, 12, 16]. Such new responsibilities can lead to stronger exposure to the public sphere, as is the case with decision-making in the Executive Committee or the General Assembly of a CFUG. This particularly applies to women living in a nuclear family without any adult son. In the absence of their husbands, these women started to attend public meetings and forest assemblies. This public exposure provided them with a new opportunity for learning and information sharing. With it, their interest in the management of the CFUG increased. This public exposure also provided them with enhanced negotiation skills and allowed them to voice their concerns related to forest management, thereby influencing decision-making.

However, in extended families, the responsibilities of the man who had outmigrated were taken up by another male member of the family, e.g. a father-in-law or brother-in-law. Thus, in both CFUGs, left-behind women who lived in extended families participated less in forest meetings and assemblies, compared to those living in nuclear families. These results are congruent with other studies that analyze gender relations within households [11, 13].

All the left-behind women reported that their husband used to be their major source of information about issues in the public sphere, e.g. the time and location of CFUG meetings and decisions taken in assemblies. When their husbands left, they lost this prime source of information. Whereas women in joint families relied mostly on other family members (male or female) to obtain such information, women in nuclear families relied mostly on neighbours and relatives. However, if the left-behind women in nuclear families were not satisfied with the information provided, they had a strong incentive to attend the next meetings themselves.

Research indicates that left-behind women tend to have a high workload [12, 16]. In the focus groups, although the left-behind women reported that their workload had increased, it did not hamper their participation in the management of the community forest. Indeed, the women noted that they were happy to attend forest meetings and General Assemblies as such meetings provided them new avenues for learning, thereby supporting their self-development.

Another issue is the remittances that outmigrated men send home and the control over this new resource. In extended families, it is mostly the male member of the family who handles the remittances. Women's opinion on the use of remittances is heard, even if they often end up being used to purchase land or to build a house. In nuclear families, usually the left-behind woman shares decision making with her outmigrated husband and thus, has more influence on the use of the remittances. Some families, both extended and nuclear, of the women in the focus groups have invested a part of the remittances to purchase alternative sources of energy, e.g. gober gas. In these cases, the remittances helped to reduce the women's dependency on forest resources, especially fuelwood.

3.3 Impact of women's engagement in the management of the community forest

Women in both CFUGs perceived that their increased involvement yielded many benefits. The forest is now better protected and forest condition has also improved in terms of forest regeneration and reported thefts of timber from the forest. Women now have easier access to forest products such as fuelwood, fodder, grass and bedding material from their own community forest. Also, women's active involvement in the CFUG has helped to draw attention to women's concerns and identify possible solutions to address them. Indeed, now that women take part in the meetings, they can voice their ideas and influence the decisions. Women are also better able to ensure that the funds generated in the CFUG are used to address their livelihood issues. Moreover, participation in the CFUG has exposed the women to public meetings and speaking in public. Successfully meeting this challenge has increased women's self-esteem and self-confidence.

3.4 Constraints and challenges to women's engagement

Despite women's active engagement in the management of the community forest, women still feel hampered by their low education level and their lack of knowledge about legal and financial aspect of community forest management. Most of the women in both CFUGs are illiterate or just literate. Therefore, women tend to develop a feeling that "they might do something wrong" if they undertake legal or financial management of CFUGs:

> "In one of the Executive Committee meetings, male members of the Committee were suggesting that this CFUG should be converted into a women-only Committee. They also asked my opinion about it. I felt a bit troubled wondering how women could deal with financial matters of forest management on their own. Most of us are illiterate. How could we handle the required skills to maintain the minutes and financial records?"
>
> A member of the executive committee of the Katunje Pakha CFUG

Though women fully acknowledged men's support behind their active participation in forest management, they also felt unsettled by men's desire to use the CFUG funds according to men's own interests. In Katunje Pakha, male members of the Executive Committee put the CFUG fund in a bank, despite female members' preferences to set up a revolving fund to provide 'easy loans' to needy families in the community. During the focus group discussion, women also mentioned other conflicts regarding the use of CFUG funds:

> "Once, a few men came to us and requested a grant from the CFUG fund to construct a road nearby. All the women signed to allow cutting trees from the community forest to raise about Rs. 35,000 for constructing the road. Later we came to know that only a small amount was used for road construction, the rest was used up by the men themselves. We felt cheated, but this event has made us more careful."
>
> Focus group discussion, Chande Manjuwa CFUG

4 Conclusion

Community forestry in Nepal is one of the highly acclaimed participatory programmes that aim to encourage the participation of local people, mainly women, in forest management. Yet, women's inclusion and active participation in decision-making remains as a challenge, and is often mere lip-service. However, the male outmigration, which is becoming a widespread phenomenon in the Mid-hills, could mediate social changes. This exploratory study was conducted to assess and analyze under which conditions male outmigration could lead to women's increased participation in the management of community forests.

As the cases of Chande Majuwa CFUG and Katunje Pakha CFUG indicate, male outmigration can indeed open a 'window of opportunity' for women. As women carry the prime responsibility of collecting forest products, they tend to be more concerned about sustainable forest management. Positive experiences in organisational management – e.g. of a savings group – or participation in a women's rights programme, has increased the women's confidence and self-esteem as well as the awareness of the options they have. Under these conditions, and with the men's support, women are willing to take on new challenges and seize the opportunities that can arise from male outmigration. The extent to which left-behind women become actively engaged in the management of a community forestry seems to depend to a large part on them being in a nuclear family and feeling that the information about the community forest they get from their social networks is not satisfactory.

Given the increasing rate of male outmigration in the Mid-hills of Nepal, there is a tremendous scope to encourage women's participation in community forestry. To realise this potential, further research is needed to identify the factors that foster women's participation and their interrelations.

Acknowledgements

We thank the users of Chande Majuwa and Katunje Pakha CFUG for their participation during data collection. Special thanks go to Bal Krishna Khanal, the forest ranger of Katunje range post for his initial support in CFUG identification and group discussions. We are also grateful to the Austrian Exchange Service for funding this research.

References

[1] WINROCK International. *Emerging Issues in Community Forestry in Nepal*. Kathmandu. pp. 4–8, 2002.

[2] Cornwall, A., Whose voices? Whose choices? Reflections on gender and participatory development. *World Development* **31(8)**, pp. 1325–1342, 2003.

[3] Agarwal, B., Participatory exclusions, community forestry, and gender: an analysis for South Asia and a conceptual framework. *World Development*, **29(10)**, pp. 1623–48, 2001.

[4] Buchy, M. & Subba, S., Why is community forestry a social- and gender-blind technology? The case of Nepal. *Gender, Technology and Development*, **7(3)**, pp. 313–332, 2003.

[5] Gupte, M., Participation in a gendered environment: The case of community forestry in India. *Human Ecology,* **32(3)**, pp. 365–382, 2004.

[6] Lama, A. & Buchy, M., Gender, class, caste and participation: The case of community forestry in Nepal. *Indian Journal of Gender Studies*, **9(1)**, pp. 27–42, 2002.

[7] Agarwal, B., Conceptualizing environmental collective action: why gender matters. *Cambridge Journal of Economics*, **24**, pp. 283–310, 2000.

[8] Lachapelle, P.R., Patrick, D.S. & McCool, S.F., Access to power or genuine empowerment? An analysis of three community forest groups in Nepal. *Human Ecology Review*, **11(1)**, pp. 1–12, 2004.

[9] Upadhyay, B., Women and natural resource management: illustrations from India and Nepal. *Natural Resource Forum*, **29(3)**, pp. 224–232, 2005.

[10] Hadi, A., International migration and the change of women's position among the left-behind in rural Bangladesh. *International Journal of Population Geography*, **7(1)**, pp. 53–61, 2001.

[11] Zachariah, K.C. & Rajan, S.I., Gender dimensions of migration in Kerala: macro and micro evidence. *Asia-Pacific Population Journal*, **16(3)**, pp. 47–69, 2001.

[12] Thelma, P., Singh, A., Luis, J. & Hossain, M. Labour outmigration, livelihood of rice farming households and women left behind: A case study in eastern Uttar Pradesh. *Economic and Political Weekly,* **40(25)**, pp. 2522–2529, 2005.

[13] Kaspar, H., I am the head of the household now: The impacts of outmigration for labour on gender hierarchies in Nepal (Part II). *Gender and Sustainable Development: Case Studies from NCCR North-South*, ed. S. Premchander & C. Müller, Geographica Bernensia: Bern, pp. 285–304, 2006.

[14] District Development Committee profile Kavre, 2007.DDC: Kavre, Nepal.

[15] Central Bureau of Statistics 2001. *National Population Census 2001*. Kathmandu, Nepal.

[16] Gurung, B. & Gurung P., Addressing food scarcity in marginalized mountain environments. *Mountain Research and Development,* **22(3)**, pp. 240–247, 2002.

The marketing of apples produced in Chihuahua, Mexico against the American Giant: a case of dumping

T. de J. Perez-Chavez, E. G. Anchondo-Aguirre,
J. L. Coronado-Quintana & M. A. Paredes-Aguirre
Full time Professors, College of Business and Management,
Autonomous University of Chihuahua, Mexico

Abstract

The State of Chihuahua is the largest producer of apples with about 50% of Mexican production. The objective of this paper is to discuss the procedure followed up by Mexican authorities who brought this issue up to international law as a case of dumping and to examine the implications of apple marketing in Mexico. In 1992, the Mexican apple growing union with headquarters in Chihuahua, detected apple importations coming from the United States of America at a price far below the price charged in the domestic market. As a consequence, the Mexican Secretariat of Commerce and Industrial Foment (SECOFI) initiated a dumping investigation in 1997 regarding the trade of apple imported from the United States of America into Mexico. The International Trade Commission reviewed the Mexican petition and agreed that dumping was made in apple importations. In August 2002, Mexico's SECOFI announced its decision to cancel the 1998 U.S./Mexico apple dumping suspension agreement. Therefore, the SECOFI resumed the anti-dumping investigation, which began in 1997. The analysis concerning the apple marketing is vital to apple growers in Mexico as a way to understand their rights and to be aware of existing procedures that ensure fair competition.

Keywords: dumping, anti-dumping, Chihuahua, Mexico, apple.

1 Introduction

Unfair practices in international commerce can affect producers in several countries, as is the case of dumping. The term "dumping" identifies the action through which a business sells a product to a foreign country at a price that is less than what it is sold for in its country of origin [1]. For example, the United States Department of Commerce sustained the case of apple concentrates imported from China, which were offered at 52% less the cost of production in their country [2]. Recently, the US-Department of Commerce initiated an anti-dumping investigation concerning light paper from China, Germany and Korea [3] alleging dumping percentages of 108% for China, 29%–75% for Germany and 40%–65% for Korea.

In the state of Chihuahua, Mexico, UNIFRUT (Union Agricola Regional de Fruticultores or, the Regional Agricultural Union of Fruit Producers) which is a Civil Association has detected dumping-like imports of apples since 1992. These imports into Mexico originated in the United States with prices clearly inferior to their normal value in their domestic market. In addition, UNIFRUT presented a petition before SECOFI, (Secretaria de Comercio y Fomento Industrial or, the Secretary of Commerce and Industrial Foment) so that an investigation into dumping could be conducted [4]. This way, compensatory dues over the import of this product can be controlled [5, 6]. This action by UNIFRUT was important and necessary when considering that Chihuahua exports approximately 60% of apples produced in Mexico [7] and that their product is excellent in terms of quality, color and flavor.

The apple production in Chihuahua is a consolidated industry that generates extensive manual labor and is essential for the state economy. At present, there is no published information over the mechanism and preceding practices of dumping that were introduced in the case of apples produced in Mexico. The recent study demonstrates that fresh apple exports from the United States into Mexico exercises dumping-like activity. Another objective was to analyze the potential damage to the national production of apples. The results allow producers in established states such as Chihuahua, Puebla, Zacatecas, Durango, Queretaro, Coahuila and Nuevo Leon to know that they are protected by international laws and that the Mexican government can help them legally in cases of dumping.

2 Learned legal actions

UNIFRUT presented before SECOFI an estimated margin of discriminatory price for the fraction of tariffs or customs 0808.10.01 that appears on the tariff of the General Import Tax Law. Because the products classified under these fractions are not homogenous but different, UNIFRUT proceeded to define the product codes considering two criteria; apple type and apple size using the Golden Delicious and Red Delicious varieties. After combining several criteria, 14 product codes were identified and the normal value calculated based on the domestic market price in the United States. The product price was obtained by weekly quotes published in the Federal State Marker News. Likewise, UNIFRUT established the normal weekly value for each product code during the investigation period. The SECOFI admitted that the method used by UNIFRUT utilized the estimate of normal value since consideration was congruent with the established depositions in article 31 of the Exterior Commerce Law (Ley de Comercio Exterior, 2005) in number 2.1 that specifies the relative agreement to the application of Article VI of the General Agreement over Tariff Customs and Commerce (Articulo VI de Acuerdo General sobre Aranceles Aduaneros y Comercio, AGAAC, 1994) as well as article 39 of the same law that defines the damage that can be caused to the national industry. Based on the pediments of importation, SECOFI found revenue of Golden Delicious and Red Delicious varieties similar to national apples. Several fresh varieties are used for human consumption.

3 Repercussions in the national market

The petition that UNIFRUT presented to SECOFI explained that the national production of apples was being constant with time (Figure 1); nevertheless, the price of apples reached the lowest level or the price increments were insignificant. Figure 1 notoriously shows that in 1995 Chihuahua produced about 11.5 million (box of 20 kg) apples and this production has been more or less constant with time. UNIFRUT explained that the low price level was due to imported apples which did not allow major increases. Therefore, apple importation has a direct consequence that damaged national producers. Clearly, the presence of low cost imports distorts behavior in the national market [5]. Moreover, the Figure 1 shows the price of the Red Delicious and Golden Delicious apples in the same period (1992–1998) where it is evident a low price in 1995.

4 Dumping and dumping margin

Dumping is when one country's products are introduced to another country's economy at a price far below the normal price in the country of origin [1]. The normal value of the product is the price offered in the domestic country. In the event that the normal value cannot be obtained, two actions can be taken; to consider the price at which the product is offered in a third country and/or estimate the production cost in the country of origin. Aside from normal value for a dumping analysis, other factors considered are product quality and the credit and conditions of the sale [1].

It is interesting to note that the practice of dumping per se, should not be considered immoral or illegal since producers can offer their products at different prices and in different markets. Nevertheless, this practice should be condemned if it threatens or causes material damage to the established industry in a country. The elements of dumping are a product being sold in a country at a lower cost than its country of origin, the potential material damage to the domestic product and the causal alliance between both events. On the other hand, anti-dumping measures should justify and exercise action only when any of these two elements unite. The dumping margin refers to differences between normal product value and the price of exportation of said product. This margin is established under two schemes; the first one is considering the average normal values in comparison with the average price of exportation, while the second is analyzed through the comparison of normal values with exportation prices obtained from realized transactions. The dumping margin is normally expressed as a percentage of the export price.

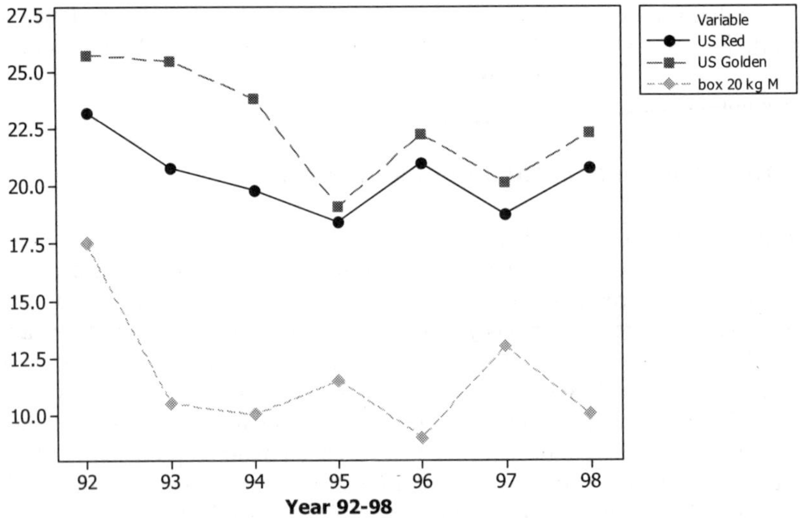

Figure 1: Apple production and price of two apple varieties in the State of Chihuahua, Mexico (1992–1998).

The World Trade Organization (WTO) condemns and restricts unfair and unjust commercial practices which generally take the form of dumping or subsidiaries. Dumping represents a fundamental distortion of basic international commerce. With exception to other obligations of GATT, they authorize a country to impose anti-dumping taxes to compensate for this unfair practice if it has caused material damage to the domestic industry. Such taxes are imposed based on the dumping margin. Although the concept of dumping is relatively simple, very complex computations are needed to adjust the factors that affect export price and normal value to make two comparable prices.

It is important to mention that Mexico just lost a dispute regarding "zeroing" with the United States. Zeroing describes analyzing the two potential stages of a dumping analysis [8]. In the first stage, a product's price in the country of origin is lower than the market price, which constitutes dumping. In the second stage, higher market prices can be specified in a country of origin at the market and consequently, dumping is not confirmed. In the case of non dumping, the commerce would assign a value of zero to this comparison, which explains zeroing, which does not utilize negative numbers.

Recently the tendency towards globalization has forced unilateral and multilateral countries ways to decrease the customary standards of imports. This induces producers to be competitive with products of other countries and consumer preference. Yet, on occasion these policies have motivated the execution of unfair practices of some countries in which the majority of cases are governed by the state, as in the case of China where costs are subsidized on a grand scale.

The financial and economic impact of dumping is evaluated when there is a reduction of product, lost sales, reduced utilities, market loss, decreased productivity, a reduced turn of investment, negative price effects and other aspects.

5 Anti-dumping legislation

Every country has tried to counteract these unfair practices in their policies with anti-dumping legislation. These measures strive to protect their domestic and regional industries and therefore, their national producers. New York apple growers in the United States lost about 81 million dollars through unfair commercial practices with apple juice which originated in China in 2002 [9]. It is evident that developed countries benefit the most at the cost of poor or underdeveloped countries. This is true because the developed countries introduce their products at less cost and consequently at a low price, due to their economic scale. On the other hand, damaged countries impose special custom barriers (tariffs) on imported products, establish some payment for exportation rights and create the payment of national subsidies, all this trying to protect their national economy.

It is a well documented fact that the United States supports their agricultural growers with high subsidiaries. This brings about high production in their country of origin and dumping in poor countries or those less competitive. In relation to market access, the gain of agricultural liberty is due in large part to financial intermediaries and not to poor countries or growers. Yet larger, developing countries can win. In spite of being under pressure, northern countries still insist upon their subsidiary regimens. While producers benefit from taxes against foreign competitors that proportion similar products at a reduced cost, consumers are forced to buy products at a higher internal price. As such, anti-dumping measures impose one of the highest costs of welfare of either commercial measure in the United States. Since Canadian producers frequently use this measure, there is no reason to believe the impact of Canada is any different. Given these significant implications to welfare, creative solutions to reduce the frequency of anti-dumping measures should be realized and can be executed with an order to prevent additional welfare loss.

Even though anti-dumping rights have a negative economic effect, this measure continues to be the preferred mechanism of protection of national industry in many industrialized countries. Traditionally, the determination of damage and the right has been a three part process; the participation of denouncing national production, the infringing of foreign products and the government. The Canadian government in its only anti-dumping deposition has special import measures that permit the consideration of public interest after duties have been imposed. This consideration to public interest could allow mitigation or suppression of functions.

6 Dumping regulation according to free commerce

Historically, this disposition has been used in limited circumstances and has not been translated to the overwhelming decrease of anti-dumping rights. The limited use of this disposition is due to political economic theory, particularly the support of production and tariffs in the form of functions. Through an economic political analysis, an interested party would have to demonstrate public interest in the Canadian tribunal of international commerce in favor of producers through a series of financial avenues and proceedings. Yet, the economic political analysis also suggests that many consumers have the power to persuade the tribunal to diminish the rights in defense of their interests with greater energy.

In the United States, the anti-dumping administration is run by the International Trade Administration (ITA) of the Department of Commerce and for the International Trade Commission (ITC). At the first level, the ITA determines if dumping has occurred through analysis of the products that have been sold in the United States at a less than normal price. In those cases in which market economies are implicated, ITA uses the price of the country of origin to determine the product normal value. This is the fair market value of the product or similar products in the domestic market of the foreign company. Once ITA determines that dumping has occurred, it first establishes a dumping margin adding the normal value price of exportation.

In the latter part of the analysis, ITA determines if the national or local industry has suffered material damage as a result of dumping. For this purpose, ITA sends questionnaires to those industry companies to collect information pertaining to damage or other pertinent information. If ITA's resolution over damages is affirmative, they in turn continue their preliminary investigation concerning dumping by sending questionnaires to all other interested parties including foreign exports and domestic imports [10]. These questionnaires collect detailed information concerning prices as well as additional information that can be used to adjust them since the export price and the country of origin are compatible. In this manner, if ITA determines based on those questionnaires that dumping occurred, ITA will amplify the investigation over the damage issue. Furthermore, in the case ITA confirms the existence of material damage; it makes the final decision over the issue, establishes the dumping margin and imposes anti-dumping taxes according to all these manoeuvres.

7 One decision, the first of its type, of a Federal Court of appeals

The United States presented various cases of dumping during the 1970s. Japan was accused of dumping steel and televisions while European car makers were also accused of dumping in the sale of automobiles. To avoid grave compensations, the majority of these producers decided to increase their prices. Subsequently, the United States adopted to retaliate by dumping semiconductor circuits on behalf of the European car makers. The most recent case also exists in the United States in respect to accusations against Mexico and Venezuela over petroleum dumping, which incurred many conflicts to determine that effectively, there had been damage to production.

8 Concentrated orange juice: a case

There are four principal producers of orange juice in Brazil; two are of Brazilian capital, one is of European and the other American. Orange juice is not consumed in Brazil, only fresh oranges are eaten which is why the majority of orange juice is destined for export. Until 1980, the demand for orange juice in the United States was satisfied wholly by national producers. But beginning in the 1980s, when there were a few bad harvests in Florida, the main orange producer, national producers were unable to cover the demand. This presented Brazil with an opportunity to export half of their orange juice production to the United States. A group of Florida producers presented an anti-dumping demand. Once the investigation was initiated, the growers realized that since Brazil did not consume orange juice, there was no presentative price. Pricing in a third country was not available either because Europe, another grand consumer, and the United States utilize the same pricing due to arbitration. A cost then had to be determined which included the price of the oranges, the costs of the industrial process plus a reasonable, beneficial margin.

It is important to note that two national drink producers, Proctor & Gamble and Coca-Cola, opposed the petition since they were large consumers of orange juice concentrate which is used in the production of their drinks. These companies considered that an anti-dumping right would elevate their costs. The decision was difficult because the growers had to be favored without excessively damaging the drink producers. Both objectives were obtained.

9 The UNIFRUT

Few sectors are as organized and forceful as the fruit growers of UNIFRUT-Chihuahua, Mexico. This organization represents 20 local associations that have approximately 2,500 producers that control an area of an estimated 30,000 ha of apple trees. In 2004, they produced nearly 18 million boxes of apples (20 kg) that represented two thirds part of the total production of the entire country [11]. As a result of the judgment of dumping against North American apples that arrive in Mexico invoiced below production costs, in September 1997 they obtained a compensatory quota of 101%. This obligated North American exporters to propose a price compromise based on $11.46 US dollars for each 42 pounds box. The Mexican government and UNIFRUT accepted the compromise but then North American growers did not fulfil their promise by selling the imports at below the established price. It would be irresponsible not to mention that the United States being Mexico's principal commercial partner, produced in 2007 approximately 155 million boxes of apples, each weighing 42 lb [4, 12].

UNIFRUT demonstrated before the Secretary of Economy that the price compromise did not represent the "packed" price for which two protective policies were interposed. Yet the Secretary did not relinquish their protection of

American exporters. Finally, the verdict in the dumping process was won by the Mexican producers and since August 2002 a compensatory quota has been established of 46.58% for the most common varieties imported from the United States.

Armed with their triumph, apple growers in Chihuahua achieved high investments to modernize their production. Protective hail mesh was installed and they increased their refrigeration capacity in a controlled atmosphere exceeding their previous tonnage from 40,000 to 112,000. A good part of the investment was possible with the support of Allianza para el Campo (the Farm Alliance), a government dominated program.

The apple business in Chihuahua employs approximately 12,000 permanent employees and about 2 million additional laborers per year. In addition, they continue their juridical bouts in dynamic form; as in the revision of the compensatory annual, three neutral judges before the Fiscal Tribunal of the Federation and a panel before North American Free Trade Agreement (NAFTA; TLCAN in Spanish).

In response to UNIFRUT's actions, American exporters and Mexican importers have not stood idly by. They interposed 30 protections, of which 25 have resulted favorably to UNIFRUT. We should specify that one of those protections was lost by errors of the Secretary of the Economy. Nevertheless, in September 2004, the American exporters once again proposed a price conference like the one established in 1998. UNIFRUT rejected the idea based upon past experience; there was no sincerity among the Americans and lack of mechanisms within the Mexican government to meet the agreement.

10 Conclusion

The practice of Mexican apple dumping by the United States market affected producers' interest and the established industry. The decision by the international tribunals guarantees impartiality to protect the producers' interests, many of whom live in countries of varying degrees of development.

Acknowledgements

The authors would like to express appreciation to the Regional Agriculture Union of Fruit Producers (UNIFRUT-Chihuahua) for the information provided. In addition, we are very grateful to the College of Business and Management of the Autonomous University of Chihuahua whose staff and General Director actively and significant contributed to this manuscript.

References

[1] Owen, V.CH.Jr. 1999. An introduction to trade remedies available under U.S. Law. Wiley Rein http://www.wileyrein.com/publication.cfm?publication_id=8009

[2] AP, 2000. Associated Press. 2000. Ruling: US apple producers harmed; (5 edition). Journal of Commerce, New York, May 17, 2000

[3] DC, 2007. Department of Commerce. International Trade Administration. Fact Sheet. http://ia.ita.doc.gov./download/factsheets/factsheets-lwtp-init-103007.pdf

[4] DiBenedetto, B. 2006. Growers fight apple export barriers; The Journal of Commerce Online Edition. Journal of Commerce. Retrieved January 16, 2008 from ABI/INFORM Global database. Document ID: 1127802721

[5] DOF, 1997. Diario Oficial de la Federación. Mexico, D.F

[6] USDA, 1997. Mexico; Antidumping investigation against U.S. apples. USDA Agricultural Trade Reports. Retrieved January 17, 2008 from ABI/INFOR; Trade & Industry database. Documents ID: 104976312

[7] SAGARPA, 1995. Estadisticas de la Secretaria de Agricultura, Ganaderia, Desarrollo Rural, Pesca y Alimentación. Mexico DF

[8] Schwab, S.C. 2007. United States Wins WTO "zeroing" disputes with Mexico. Office of the United States Trade Representative. http://www.ustr.gov/Document_Library/Press_Releases/2007/December/United_States_Wins_WTO_Zeroing_Dispute_with_Mexico.html

[9] Schumer, Ch.E. 2002. Schumer: New US commerce dept plan could cost NY apple growers millions of Dollars. United States Senate http:www.senate.gov/schumer/schumerwebsite/pressroom/press_release/PRO1197.html

[10] DiBenedetto, B. 2005. Mexico lifts apple tariff; The Journal of Commerce Online Edition. Journal of Commerce. Retrieved January 17, 2008, From ABI/FORM Global database. Document ID: 851243971

[11] UNIFRUT, 2007. Union Regional de Fruticultores del Estado de Chihuahua. Estadisticas basicas del estado

[12] WHT. 2002. World Horticultural Trade & US export opportunities http://ffas.usda.gov/htp/Hort_Circular/2002/02-11/Stats/NORTHERN%20H.pdf

Author Index